FIRE SCENE DISCIPLINE

FIRE SCENE DISCIPLINE
TRUE STORIES OF CLOSE CALLS

Danny Sheridan

Fire Engineering®

BOOKS & VIDEOS

Copyright © 2024 by
Fire Engineering Books & Videos
110 S. Hartford Ave., Suite 200
Tulsa, Oklahoma 74120 USA

800.752.9764
+1.918.831.9421
info@fireengineeringbooks.com
www.FireEngineeringBooks.com

Executive Vice President: Eric Schlett
Vice President, Group Publishing: Amanda Champion
Acquisitions: David Rhodes, Diane Rothschild
Director, eLearning and Books: Starlet Franz
Production Manager: Tony Quinn
Developmental Editor: Chris Barton
Operations Manager: Holly Fournier
Cover Designer: Brandon Ash
Book Designer: Robert Kern, TIPS Publishing Services, Carrboro, NC

Library of Congress Cataloging-in-Publication Data Available on Request

ISBN13: 9781593705862

Printed in the United States of America

1 2 3 4 5   27 26 25 24

Disclaimer

Readers can anticipate and experience a sense of danger. How does a firefighter evolve in a career from a probationary office to a company officer? How does a firefighter learn and survive the job each day? This personal perspective tackles fire attack, ventilation, rescue, building collapse, communication, and service to others. The stories here may jump quickly from one action to the next with a sense of what a firefighter's experience is on the fireground. Some names have been changed.

Contents

Introduction

It's important to learn from our mistakes, but it's even better to learn from someone else's mistakes. In this business of firefighting, it is impossible *not* to make mistakes. Firefighters are charged with the unenviable task of solving a situation that can rage out of control in a matter of minutes. In the business world, managers take weeks if not months to plan for an event. Firefighters are given minutes.

Firefighters can resolve most incidents successfully. We do so because we plan, train, and rely on our experience from previous fires and emergencies to attain a positive outcome. Each fire or emergency that we respond to goes into our own personal slide deck—visual training aid—to be called upon in the future when we are faced with similar situations. It doesn't matter whether we were actually there; what is important is that the lessons learned are retained and disseminated.

Modern Fires Versus Legacy Fires

Fires today are tremendously different because of fuel packages, open living spaces, insulation, and the size of homes. These are some of the factors that contribute to the *modern fire*. However, the fires in this book differ from some of the author's earlier fires since they behaved in the way of a modern fire before that term was used in the fire service.

The following stories will add more information to your visual aids or slide deck.

It's not critical to each of these stories as to what names are used or where they happened; the important message is to convey a lesson learned that can be passed on to another firefighter.

So much has changed since the author entered the fire service in 1986. Fires absolutely behaved differently back then, but today we refer to them as *legacy fires*. The rules that applied at the time may not apply today. For example, the mantra in the fire service used to be to "never open the nozzle on smoke." We know today that this is an antiquated notion and needs to go the way of horse-drawn carriages.

One of the most impactful changes over the decades has been the introduction of bunker gear. In one of his earliest experiences, the author and his partner were on the floor above the fire in a Type 5 private dwelling. With a lack of experience, they didn't realize the heat they felt was *not normal*. They assumed that fires are supposed to be extremely hot. When the room flashed over, they were fortunate to be near a window and dove downward on a portable ladder. Lessons learned: the importance of portable ladders *and* recognizing extreme heat.

In another fire—pre-bunker gear in the 1980s—the author and his partner were on the floor above in a Type 3 vacant five-story multiple dwelling. They found themselves on a hoseline that had been vacated and proceeded to attempt to extinguish a floor that was totally involved. It was so hot that they could not kneel but rather were forced into *duckwalking*. Eventually, they were surrounded by fire on all sides when they lost water. It turned out that the chief ordered a tower ladder to be put into operation, but they never received that message. The tormentor poles used to maneuver a 75 ft. ladder were put down on top of their hoseline.

Communication is one of the first lessons learned here. Then, had the two of them been wearing bunker gear, they may not have realized the gravity of the situation. The fire service experienced growing pains with the new bunker gear. Lots had changed: ladder companies now had better forcible-entry tools, specifically the Hydra Ram. They were getting into the occupancies faster and deeper while the engine firefighters were not at the door with the hoseline.

It used to be that the ladder company would still be forcing the door while the engine waited with the hoseline at the front door, potentially creating a recipe for disaster because the firefighters no longer felt the heat like they normally would have without the hoods and bunker pants; they would be entering a very hostile environment, unbeknownst to them.

The playing field was leveled again a few years later (approximately 1997) with the introduction of the thermal imaging camera (TIC). Firefighters were now given the ability to use the camera to check the conditions that were present despite not having the ability to feel anything.

In one story the author decided to conduct an experiment. It may sound ridiculous these days, but understand that the author spent many years without having the gear and still was nervous about being completely encapsulated. His crew pulled up to fire out the windows in a five-story Type 3 multiple dwelling. He was a lieutenant working in the engine, so he did what he was taught by one of the senior lieutenants from the engine where he was a firefighter in the ladder. He watched as the lieutenant disappeared into the floor below to get the layout of the fire apartment. Layouts are the same on every floor in these types of occupancies.

He got to the third-floor front apartment, which was directly below the fire apartment on the fourth floor. After checking out the apartment he made a mental note: "Two rooms in the front, the rest in the rear." He decided since there was a heavy volume of fire to totally encapsulate himself; he knew the layout and knew where the fire was, so all that was needed now was the hoseline and his guys would knock the heck out of it. He met his team on the fourth floor, called for water, and the nozzle firefighter bled the line. They entered the apartment and proceeded to extinguish the two rooms that were burning. He thought to himself, "This is amazing, I don't feel a thing." They extinguished the fire very quickly, and he decided to get the team relieved. He kept his gear on until he got to the street. He never felt a thing.

It was during this time when he was promoted to lieutenant when the author really started noticing the changes on the fireground. He had spent his last 4 years as a firefighter in the "squad," so he wasn't really in the front row anymore. When he was new to the ladder company in 1986, besides not knowing anything about what was normal and not normal, he always was on the inside team. The ladder company had two teams, one inside with the lieutenant or captain, and the other with more experienced firefighters on the outside positions. Fires grew slower and took longer to flash when he first started. Residents in the city back then didn't have as much, apartments were smaller, materials were cellulose based (nontoxic and biodegradable), and the buildings themselves were leaky. Besides the obvious arson fires, which were always apparent, the fires did not behave as violently as they did in present time.

Ventilation-Caused Fire Behavior

The only time fires behaved erratically was when a firefighter vented inappropriately. In one story here, the author got caught behind a fireball in a high-rise when a roof firefighter took the windows without first checking in with the firefighters on the fire floor. A similar event occurred in another story, but with a different outcome. Miraculously, the wind was blowing into the apartment from the hallway as opposed to the other way around. A firefighter took a window in the hallway without checking in with anyone and created a fireball that blew out the windows and engulfed the whole apartment. Had the wind been blowing in, they would have been incinerated.

Basement Fire Close Call

A fire in the basement of a brownstone almost ended in tragedy (we've seen similar incidents in Massachusetts and Maryland as well). A door was inadvertently opened in the rear, creating a flow path and flashover that nearly killed the inside team. Although the firefighter that opened the door created this situation, at the time he didn't perceive the danger. He was doing what he always did, vent, enter, and search (VES). What was missed was that it was not coordinated with the firefighters on the inside.

The author also addresses incidents when things were within the crew's control. The message is to take ownership for *your* part, especially when facing an ever-changing fireground. We can't always blame our poor performance on the changing fire environment. In some stories, the author was being too lax, making assumptions that when he reached the fire floor he would be okay. But what about taking the time to do a proper size-up like he was taught by his lieutenant, who used to ask him, "What floor were we on?" to which the lieutenant received a dumbfounded look?

In one of the following fire stories the chief asked him a very critical question: "Is the fire in the cockloft?" This is important because if it was, then the chief would definitely need to transmit an additional alarm. The author didn't have the answer; the best he could do was tell the chief, "If I'm on the top floor, then yes." I'm almost certain that the chief was not happy with that answer; I know I wouldn't be. Another time he brought his inside team to the fire floor when he should have been on floor above. It was quite an embarrassment when he bumped into the first truck in the fire apartment. It's almost as if he could hear his wife's voice telling him to remember the five "P"s—*proper planning prevents poor performance.*

As a lieutenant, the author was always in the front row at the seat of the fire. He began noticing that the smoke was becoming much thicker and hotter. When he was a firefighter carrying the 2½ gal extinguisher, he remembers the smoke not being as violent. At one particular fire in a high-rise fireproof multiple dwelling where the door was opened, he couldn't see the hand in front of his face, and he didn't feel anything as well because he had his hood up. The assumption was, based on his experience, that there were a good few rooms burning. After searching for what seemed like an eternity, the fire could not be found. It was later determined that the fire was in a desktop computer that had totally burned down.

Upon his promotion to captain in 2003, the author noticed that the fireground was changing; even though he had no official studies to review, he noticed changes. An apparent simple rubbish fire nearly incinerated apparatus and burned three firefighters on the back step. The garbage was burning a plastic garbage can that had been recycled from the Bronx Zoo. Unbeknownst to anyone, the 55 gal drum had previously held soap that was used to wash the elephants. Firefighters used the water extinguisher to attempt to put out the fire while the author remained in the cab to witness a fireball that erupted from the can that reached 20 ft. high. Later, it was found that the garbage had melted and formed a pool of flammable liquid, which ignited. The scary thing is that these same materials are now found in regular household furnishing.

From these many lessons, the author has seen over his 36 years a gradual change but none that was more noticeable than a fire that occurred in 2006. Firefighters would innocuously tout their accomplishments by the condition of their gear. Back then it was a badge of honor to have a "salty" coat or helmet. Other firefighters would subconsciously judge a firefighter based on what condition their gear was in, particularly the front piece of the helmet. I myself will admit I took pride in my front piece being blackened. I remember after a particularly nasty fire in a high-rise multiple dwelling, the senior fire fighter sitting across from me swiped my helmet and said, "You're getting pretty salty, kid" (the smudge is still on my front piece to this day).

In one story, the author is the captain working in a very busy engine company, when they receive a report of a fire in a six-story non-fireproof multiple dwelling. The chaos aside, which you will read about, one important component was the voracity of the fire.

The way the fire was pushing out the windows was something I'd only seen in the numerous vacant building arson fires I had been to in the 1980s. The heat coming out of the fire apartment was unbearable. The ladder company inside team was composed of a very seasoned well-respected lieutenant along with a brand-new probie. When they exited the apartment, the probie's gear looked as if he had been on the department 20 years. (My front piece was brand new as well and it also looked like it had been thrown into the fire.) This is a testimony to what I perceive as the change in the fireground. I see firefighters today with front pieces that are indiscernible. In the War Years of the FDNY, firefighters sometimes went to seven fires a tour and their front pieces didn't look that beat up.

The author had been acting as a battalion chief since 2004 and was officially promoted to battalion chief in 2008. All his experiences in his previous ranks came into the forefront. Watching and learning from some of the greatest firefighters, officers, and chiefs to ever suit up, he was lucky when he came on the department to learn from so much experience. Lieutenants had over 20 years, the captain had 30, and all the chiefs had over 30. He watched in amazement how calm, cool, and collected they were. He remembers a fire in a partially occupied five-story multiple dwelling where the fire was on all floors. There were at least five hoselines operating (each unit received a unit citation). The deputy chief and battalion chief in front of the building were not the least bit rattled. They had a tower ladder at the ready but never used it; instead they relied on the tenacity of these tremendous engine companies to extinguish the fire, which they did. It was this type of leadership that I hoped to emulate when I became an officer and eventually a chief. I took all my experiences both good and bad to make me what I am today.

One particular fire was in a vacant three-story Type 5 wood frame. The fire was on all floors and the building was marginal at best, but that didn't deter my ladder company; we were bent on getting in there and fighting the red devil. The deputy chief stood in front of the building and prohibited us from entering. Needless to say we weren't happy, especially when I was relegated to "bucket duty": we were gonna dump the tower ladder into the building. At one point the building collapsed right in front of my eyes while I was trimming the windows and cornice. I never forgot that. When later I responded to a fire in a three-story vacant frame, guess what call I made from the firehouse apparatus floor?

Collapse is a major concern of battalion chiefs, particularly in commercial buildings where there is no life hazard.

How did we miraculously survive a few collapses? Today, I pay strict attention to fires with respect to collapse, especially with all the lightweight construction going up in cities across North America. Collapse can occur quickly and usually without warning. It's imperative that all firefighters, officers, and chiefs do their part to make sure that we are not in harm's way. If you are not aware, the signs are easy to miss; I remember as a captain at a fire in a one-story taxpayer stepping over a huge chasm between the bearing wall and the roof. If it weren't for the attentiveness of one of my firefighters, I would have glanced right over it. I reported the riff to the chief and he pulled everyone out and set up the tower ladder. It's like the mindset of "it takes a village."

I've been a chief longer than I had been any other rank, and my father-in-law was right: I enjoy this rank the best because I find that the battalion chief can have the greatest impact on the success of the operation. I hope that these stories and the lessons learned will help you in any way possible. You may have a situation and think, "Hey, I read something about that, and this is what happened."

When I was brand new, my lieutenant who had 20 years on at the time told me, "You are always learning." I was amazed at that time because this fire lieutenant was someone who (I thought) knew everything. He had seen it all, but he was also advising me to remain teachable. I've often heard that if you think you know everything, you probably don't, and if you think you don't know anything, you are probably in good shape.

Glossary

can. The can position's main duties are getting into the building and rescuing trapped people. The firefighter in the can position is part of the forcible-entry team that uses a water extinguisher to attack the fire through the front door.

irons. The firefighter in the irons position is part of the forcible-entry team. Their first duty is to gain access into the building, by force if necessary, and begin a search for victims trapped by the fire.

outside vent. The outside vent position's main job is to vent the fire from the outside to create an opening for heat and smoke to escape the building. In this position the firefighters often work on fire escapes, ladders, and tower-ladder buckets. Gated windows, high fences, heavy security doors, and objects left on fire escapes make the outside vent's job more difficult.

roof. Firefighters in the roof position use tower-ladder buckets, aerial ladders, portable ladders, fire escapes, and roofs of neighboring buildings to access the fire. This job is very dangerous because the roofs of burning buildings often collapse. Their first job is to give a report of the rooftop fire conditions to the commander.

truck chauffeur. The truck chauffeur drives the ladder truck to the fire scene. At the fire, the chauffeur is responsible for raising and positioning the ladder and bucket, then removing trapped occupants from the front of buildings.

truck officer. The main jobs of the officer at the fire scene are to keep track of the company and lead them into the fire. The officer is part of the forcible-entry team that gains access and locates the area of fire so it can be put out by the engine company.

Soften the Target: An Offensive Transitional Attack or Quick Exterior Knockdown

> **Scenario:** The captain arrived on scene of multiple buildings on fire. He wanted to stop the fire in the third building on the B side, but his decision to enter the building without a hoseline nearly cost him his life.

Engine 164 was at the epicenter of the War Years in the borough. Much of the area loaded with five-story tenements was now an industrial park. The rest of the area where the vacant buildings had been was either renovated or torn down and replaced with lightweight smaller two-family private dwellings (fig. 1–1).

Danny was concerned about two old vacant buildings in the part of his response area that had not been turned into an industrial park. They were the first and

Figure 1–1. Lightweight construction private dwelling

second buildings in a row of five wood row frame buildings separated by a 5 ft. alley on the northern edge of his response area (fig. 1–2).

It was one of those borderline fire alarm boxes that, depending on which box the dispatchers transmitted, he could be first or second due. Since he had been promoted to captain, Danny went to few fires in vacant buildings.

For an engine company, the difference between first and second due is like night and day. The first-due engine gets all the glory. The second engine ensures water supply and assists with getting the first hoseline in operation. He knew it was only going to be a matter of time before they lit these buildings up. The Department of Housing and Urban Development (HUD) windows and doors had all been compromised and the drug dealers were going in and out of there on a regular basis. Occupants lit fires to keep warm during winter. It was not uncommon for the drug users to nod off and fall asleep while the fires still burned.

There had been a few minor rubbish fires in the buildings over the course of a few weeks, until one night they lit up the cellar. Engine 164 arrived simultaneously with Engine 189. Engine 189 was first due on the ticket, so Danny told them to go ahead and stretch the first line. Because the building was only two stories, they stretched a second line to back them.

The operation was routine. Engine 189 extinguished the fire quickly while Engine 164 took care of the fire that got into the walls and spread to the kitchen on the floor above. Afterward the two officers got together, and Danny gave Tim some good-natured ribbing about how they stole his first-due box. On a serious note, they both knew that this was only the beginning: whoever was setting these fires wanted the buildings destroyed. They both understood that with this type of construction, fire extension to the other three occupied buildings was going to be a problem.

Figure 1–2. A rehabilitated multiple dwelling

The next set of tours, both officers were working again. In the late afternoon, Engine 164 got relocated to cover another engine in the north part of the city. Danny was annoyed because he knew that the fire the company up north went to was nonsense. It was a minor fire, but because the company was isolated, the firehouse always needed to be covered. They had been there less than an hour when the other engine returned to quarters. Engine 164 headed back to their response area a few miles south.

Normally Danny would have the chauffeur take the highway to get to the relocation and back because it was faster. The downside was that they would be unable to respond to any calls if there were any. This day Danny had a sixth sense. He told his chauffeur to take the avenue; it would be slower and there were a few lights, but at least they would be mobile. Also it was rush hour, so the traffic on the highway was terrible and they would be stuck anyway. They started heading back to quarters when the radio on the apparatus came to life. The dispatchers were notifying his neighboring engine company that they were getting numerous calls for a fire at the location of the previous fire in the cellar of the vacant building.

"Dispatch to Engine 189, we're loading up the box; we're getting numerous calls on this."

Danny was aggravated. Not only did he get relocated for some nonsense rubbish fire, but he was now missing out on a working fire. Tim reported that he could see a massive black cloud over in the direction of the box location. Danny knew that it was the two vacant buildings that he had been worried about. He notified the dispatcher that he was available to respond. The dispatcher asked his location, and like everyone else does, he lied: "Engine 164 to dispatch, we're pretty close; put us on the box."

"Ten-four, 164 take it in."

Tim cut off Danny's transmission with a very urgent message: "Engine 189 to dispatch, transmit a second alarm; we got heavy fire in a two-and-a-half-story vacant private dwelling with extension to both exposures."

Tim's crew from Engine 189 did an incredible job of stretching a hoseline into the fire building while also using the deck gun to knock down the main body of fire on the outside of the building (fig. 1–3).

The fire had already extended into exposure D. The chauffeur played the deck pipe, alternating between the exposure and the fire building. Tim made a decision to use the 500 gallons of water in the booster tank for the deck gun as opposed to driving to a hydrant and securing a positive water source. If he had taken the hydrant, the deck pipe would not have been in position to knock down the fire. He made the right call.

Normally Engine 164 would have rolled in pretty much at the same time as Engine 189 and they would have worked together to get the deck pipe in

Figure 1–3. A fully involved blaze at a vacant private dwelling

operation, another example of how one little thing can manifest itself into bigger problems (fig. 1–4). The next engine was much farther away. The deck pipe quickly ran out of water, and the fire was able to gain a good head of steam because they were not there to supply the first engine.

Because the second alarm had been given right away, all the units were arriving at the same time. The fire had gotten into the occupied buildings on the exposure B side. When Engine 164 arrived, it was mayhem. The deputy chief ordered a tower ladder to the front of the building and the second tower ladder to the open lot in the rear.

The fire was taking off faster than the units could get hoselines stretched. Danny checked in with the chief and asked him what he needed from him. The chief frantically replied, "Cap, just pick a building and get a hoseline in there" (fig. 1–5).

He sized up the situation and figured that he would skip the building next to the fire building, exposure B, and head into exposure B2. The fire was already consuming the immediate exposure. Exposure B2 was salvageable. They found an available pumper near the front of the building and stretched the hoseline to the front door of exposure B2.

Figure 1–4. Fire quickly spreads into the exposure.

Danny told the crew to wait at the front door while he scouted out the situation. He teamed up with two other firefighters and ascended the narrow stairs. When he got to the top of the stairs on the second floor, there was nothing but orange outside the window. The bathroom door at the top of the stairs was closed, but Danny knew that there was fire in there.

At this point Danny should have returned to the first floor and had the guys bring the hose up, but he pressed on. He just violated his own rule of never passing fire without a charged hoseline, but he passed the fire and made it to the bottom of the stairs leading up to the attic.

When he got to the attic, the heat was unbearable. He was on his belly looking for the fire. He decided that he needed to turn around and get the heck out of there; he knew that he messed up and something was not right. That's when it happened!

He heard it before he saw anything: a bang, then darkness. He thought to himself that this was it, he had finally done it and pushed the envelope too far. Instinct took over. He was in survival mode, and before he knew what happened, he was already on the second floor, diving down the rest of the stairs to the first floor where he landed on his guys at the bottom of the stairs.

Figure 1–5. Fire is fully involved in the rear and spreads to all exposures.

It took a few seconds to get his bearings. Danny was shaken up pretty bad. He had been in tight spots before, but this was the closest he'd ever been to being caught in a flashover. His head was a bit clearer, and it now occurred to him that he was with two firefighters. He was about to give a Mayday when the two of them came flying down the stairs behind him. They all came out of this unscathed and were able to get out under the flames, which was nothing short of a miracle.

The building had erupted into flames. He wasn't sure what was happening, but he decided to let the tower ladders deal with it. He reported back to the command post; this now was an exterior operation (fig. 1–6).

This fire always haunted Danny. He would get uneasy when he thought about what could have happened. He was a good Catholic and felt that somehow God had intervened because he shouldn't have gotten out of that one. A bunch of years had passed, and Danny was catching up with one of his mentors, for whom he had such great respect. They were going to work on a project for the department concerning rehab buildings. As firefighters always do, they started telling fire stories.

He was telling Joe the story in great detail of how he passed the fire, the fire outside the window, and the excruciating heat on the top floor. Joe told Danny that a similar thing had happened to him once.

Figure 1–6. When the fire becomes an exterior fight (courtesy of Michael Dick)

He was working in Ladder 88 that was second due for a fire in a three-story row frame. The fire had gotten into the walls and spread up into the cockloft, unbeknownst to him and his crew. They were on the top floor, and it began getting very hot, but there was no visible fire. Then, without warning, the room flashed over. He and his crew managed to escape by diving down the stairs.

It occurred to Danny that this is what happened to him and the other two firefighters. The fire must have spread up the rear walls from the bathroom and entered the cockloft. At some point it blew down onto him the same way it had happened to Joe. Both groups were fortunate to have survived.

Lesson Learned

The problem with hidden fire, according to Francis Brannigan in *Building Construction for the Fire Service*, is that "concealed fire which bursts out of a hidden void, and lightning-like spread of fire over combustible surfaces, are equally hazardous and may well account for as many firefighter casualties as collapse."[1]

1. Francis Brannigan, *Building Collapse for the Fire Service*, 3rd ed. (Quincy, MA: National Fire Protection Association, 1992), 12.

2

Firecrackers and Fire Resistance

Scenario: The inside team risked a lot to get to a rear bedroom where there was a report of a baby trapped on the fifth floor of a five-story tenement building (a low-rise fireproof multiple dwelling).

First Real Fire

Danny remembered how he waited patiently to go to his first *real* fire. He'd only been to a few minor calls—a burning mattress in the stairway and a few compactor chute fires. On a chilly day tour in October, the day finally arrived. Ladder 85 responded to a report of a fire on the third floor of a five-story Old Law Tenement. As they approached, the building flames were visible shooting out the windows on the B side. They entered the lobby, making their way past the screaming occupants up the interior stairs. Danny's knees didn't stop knocking together, and his nervousness changed to almost pure terror when they arrived on the fire floor. Thick, dark black smoke was pumping out of the wide-open apartment door. It looked like his first real fire had finally come (fig. 2–1).

Lieutenant Gill, a 30-year veteran in the department, quickly donned his self-contained breathing apparatus (SCBA) and disappeared into the darkness. However, Danny fumbled to don his SCBA, acting like a fish out of water. The fire was in a bedroom down a long hallway on the right side of the apartment. He treaded very lightly through the threshold into the oven-like conditions in the hallway. He was able to make out a faint glow through the facepiece of his SCBA as the lieutenant yelled to Danny, "Get the water can down here" (fig. 2–2).

"Is he talking to me?" Danny thought. "Does he really mean that he wants me to physically come there with the water can? That's insane." He was reminded

9

Figure 2–1. Smoke pouring out apartment door. The ladder company inside team (officer plus forcible-entry and extinguisher firefighters) is responsible for entering the apartment to locate, confine, and if possible, extinguish the fire.

Figure 2–2. The "can man" carries a 2½ gal extinguisher along with a 6 ft. hook, which comes in handy when the door needs to be closed and it is impossible to get near enough to close it.

about the joke he heard in probie (probationary firefighter) school about the lieutenant who called out in the darkness for the water can. A few seconds later the water can came flying down the hallway. He briefly thought that it wouldn't be too funny if he flung the water can down the hallway.

He mustered up enough courage to crawl down this long dark hallway toward the fire (fig. 2–3).

The glow became more visible as he got closer, not to mention that it was getting hot really quick. The fire burned out into the hallway and he couldn't get to the door to close it.

He somehow made it to the door when he remembered the technique he was taught at the academy. He squeezed the trigger and opened the nozzle up into the raging fire. He put his right index finger over the nozzle to make better use of the 2½ gal of water in the extinguisher. He couldn't believe how much of the fire was being knocked down. It died down enough that they were able to close the door and confine it. After the water can was expelled, they vented and searched the rest of the apartment. Still a bit nervous, he was glad the worst was over. He had faced his fear and overcome it. Now, the lieutenant ordered him to vent: "Hey kid, take those windows in the living room."

Charged up with adrenaline, he nearly decapitated the forcible-entry firefighter with his 6 ft. hook while smashing the window.

─────

Figure 2-3. Hallway in a tenement building

Tower Ladder 85 responded to vacant building fires on a regular basis (fig. 2–4). Vacant building fires enabled his company not only to get to know how to work together and bond but also to reconsider these dangers before they would initiate an interior attack:

- Size and intensity of the fire
- Location of fire within the structure
- The stability of the structure
- Safe access to the fire area

Danny was developing his firefighting skills and boosting his confidence. Tradesmen become good at their jobs because they do them every day; it was all about "hands on" and "muscle memory."

Danny started to feel more comfortable in his job. He felt like he actually knew what he was doing. There are two jobs to be done when it comes to operating the tower ladder: one firefighter operates the controls or "flies the bucket," but the other job was more desirable—operating the tower ladder pipe. That firefighter directs that massive 800 gpm stream into the belly of the fire. Operating the tower ladder bucket was intimidating, but he was gaining confidence and improving. He became proud of his ability to fly the bucket of the tower ladder. A senior firefighter, Big Tom, took him up with him when they had outside

Figure 2–4. What looks like a typical vacant building fire in the Bronx, circa 1986

operations in these vacant buildings. Even though Tom had over 20 years in the department, he still got a thrill using the pipe to extinguish the usually large amounts of fire. Tom's affinity for using the pipe forced Danny to learn how to operate the bucket.

After 3 years at the firehouse, Danny became a first-grade firefighter, and the officers in Tower Ladder 85 were assigning him to the forcible-entry and roof positions when he was working. This was okay with some of the senior firefighters because they enjoyed carrying the water can, especially Billy. Danny and Billy were on shift together and had become close after facing many tough fires together. Billy was not the biggest firefighter in the firehouse, but there weren't many who were tougher. He and Billy made a good team. After a while, you begin to rely on one another in life or death and, eventually, build total trust.

It's important for firefighters to get lots of experience on the inside before they commit to being an engine company chauffeur (also known as the engineer). It is of such importance that other departments require firefighters to take a competitive exam for promotion. In the FDNY, the captain will decide when they feel that a firefighter is ready to go "into the seat." That's a FDNY term for firefighters who become chauffeurs. Firefighters with experience know the importance of getting water into that first hoseline.

Fourth of July started like any other summer night with lots of running around, chasing car fires and false alarms. Many of the older guys couldn't care less about this summer holiday because they'd seen so much fire duty and would rather be home celebrating with their families. The younger guys tried to work swaps in that night because it was busy and young firefighters train for such a night.

The real "fun" starts at dusk when the neighborhood reminds you of Beirut, Lebanon, in the 1970s. Rockets and firecrackers and firearms shoot off everywhere. Around 2300 hours, after about 2 hours of nonstop chaos in the neighborhood, they received a phone alarm for a fire in a six-story low-rise fireproof multiple dwelling (fig. 2–5).

Per policy, the house watch firefighter called out the alarm, turned on all the lights, hit the foghorn, and acknowledged the run on the teleprinter before gearing up and heading to the rig. "Everybody goes, first-due phone alarm, fire on the top floor."

At the same time, there was a second alarm in a vacant building nearby; the chief at that fire was looking for tower ladders, but luckily Tower Ladder 85 was spared. They already had two tower ladders, 98 and 187, on scene. They didn't even get called to relocate to either of those companies. (When fires happen in other areas, Tower Ladder 85 will sometimes have to cover the empty firehouse.)

———

Figure 2–5. A low-rise fireproof multiple dwelling

This building was part of a housing complex. The buildings replaced a row of tenements that had been destroyed by numerous fires in the 1970s during the War Years.

The buildings were interconnected throughout a whole city block. There was a huge courtyard in the center of all the buildings; on the west side of the complex was a car park negating any access for the tower ladder (fig. 2–6).

The building had a dry standpipe system, which does not contain any water in the piping until it is supplied, unlike most of the buildings in the area that were 20-story Housing Authority buildings that all had wet systems. The benefit of a wet system is that it needs only to be augmented with water from a fire department connection (FDC). With FDCs on the north and south sides of the complex, the south being the street side, it was a tricky setup even for the most senior of chauffeurs.

As they approached the scene, the civilians in the street were directing the firefighters into the courtyard. (When people are pointing you in a certain direction, you know something's happening.) They were telling them that there may be people trapped in the apartment. Kids in the complex had been shooting fireworks into open windows and lit up this apartment on the top floor. Being Type 1 (fire resistive), the apartments had no exterior metal fire escapes; if anyone was in the back bedrooms, they were not getting out on their own (fig. 2–7).

Fire was out the windows on the sixth floor in the living room in the courtyard. The bedrooms were probably cut off and there was no access for the bucket.

Lieutenant Esposito (Espo), Billy, and Danny entered the building, located the stairs, and made their way up to the sixth floor. Taking the elevator is never

Figure 2–6. Typical building complex

Figure 2–7. A high-rise building with no rear fire escape

done when the fire is below the seventh floor. When Danny opened the door from the enclosed fireproof stairway into the hallway, it was lights out—the hallway was fully charged. Danny knew that it could only mean one thing: the door to the fire apartment was open. In Type 1 buildings, the hallway becomes part of the fire area when the door is open, meaning there is no difference between the inside of the apartment and the hallway.

The scissor stairway was tricky because when you enter the stairway, the door comes out on the opposite side on the floor above and so on. Danny was in the middle of the sixth floor. Locating the fire was going to be difficult (there weren't any thermal imaging cameras back then). You relied on your senses but mostly you followed the source of the heat. Sometimes it was your gut instinct and sometimes just dumb luck. The heat was everywhere. It bounced off the walls, like being on the inside of a pizza oven.

Danny sensed that the fire was to the right, but he did not know why; it was probably because subconsciously he noted the flames out the windows from the D side of the building. Danny crawled on his belly down the hall. Espo and Billy went left. The glow coming from the far apartment was now visible. He yelled back to the other two, "I found the fire apartment. We need the water can."

While they were looking for the fire, the engine was busy hooking up their 2½ in. hoseline on the floor below. The new chauffeur was having difficulty with the water supply, so it was going to be a while before they had water in the hoseline. Espo and Billy met Danny at the door to the fire apartment. With a wall of fire in front of them, the living room was fully involved. The bedroom Danny needed to search was on the right, and the hallway leading to the bedroom was visible through the flames.

Danny had implicit trust in Billy, who was a natural, and was comfortable when Billy carried the 2½ gal water can. They were going to have to wait for the engine because there was too much fire blocking the hallway to the rear bedroom. The neighbors insisted that the baby may still be inside. He looked in frustration at Billy, then Billy surprised him and said, "Sparky, I got this."

Billy worked like a magician, making fire disappear in the couch and chair with every ounce of water remaining in the 2½ gal extinguisher. Danny didn't hesitate and dashed quickly, using his gloved hand to cover his exposed ears past the raging living room fire, with Espo right behind him. Their lives were literally in Billy's hands.

When Danny got to the bedroom, he heard a low cry, almost like a whimper. He started a right-hand wall search and came across a crib. He felt around the crib, came across a small body, and scooped it up. With the tiny body tucked under his arm, he made his way back out to the front door, followed by the lieutenant. The fire was still burning a bit, but Billy had expelled all the water from the water can, knocking down the fire enough for Danny to escape with the body.

Danny took off his SCBA as he raced toward the stairway, Espo still on his heels. It was then that he realized the baby he thought was in the crib turned out to be a *cat*. Espo was not happy, but there was no way they could have known that the family had gotten out before they arrived. They discovered later that the kids in the courtyard had shot a bottle rocket into the window and caught the curtains on fire. They fled the apartment, leaving the door open. The cat unfortunately didn't make it.

Young Firefighters Survive Flashover

> **Scenario:** Two young firefighters were caught in a flashover on the second floor of a two-story wood frame. They survived by bailing out a properly positioned portable ladder.

There is no substitute for experience in the fire service because it is critical to a fire company's success. So it wasn't the best idea to have three probies (probationary firefighters who just passed through the fire academy) working with a covering captain on Christmas Eve. Danny Sheridan, Thomas Sullivan, and Jairo Sanchez barely had a year of experience combined and were assigned to the firehouse 3 months ago. The three probies agreed to work the holiday covering for the older guys who wanted to be home and spend the night with their families. Joe Huber was filling in for Captain Dave, working again this night as acting battalion chief.

The probies all wanted to do the right thing for the senior men, but sometimes the best of intentions put firefighters in harm's way. Three probies and a covering captain just filling in for the night can become a recipe for disaster.

Tower Ladder 85 received a phone alarm for a fire in a private dwelling at around 0400 hours. The bleary-eyed house watch firefighter called out the box: "Everybody goes, phone alarm, fire in a private dwelling, engine and truck both second due" (fig. 3–1).

With already 6 in. of snow on the ground, it wasn't stopping. The streets were thankfully empty, but it was a slow go for both companies. Snow always adds another level of difficulty to any fire scene: the cold, ice, and snow make it difficult to move quickly and put up portable ladders and the like. They were assigned second due, which meant that they were tasked with going above the fire to search. The dispatcher was receiving a few phone calls on this one; it appeared as if it was going to be a working fire.

Figure 3–1. Firefighters arrive to a fire in private dwelling (courtesy of Michael Dick).

The first-due engine was already on scene, as they were located only a few blocks from this fire. Since they were a single engine, they were quickly able to get out of quarters. The unplowed street was narrow and laden with heavy snow, making it difficult for the huge tower ladder to make its way through. In the distance they could see a group of civilians in the street (fig. 3–2). The civilians were frantically waving them into the fire on the first floor of a two-story wood frame private dwelling. The fire was out the windows in the front on the D side. One woman grabbed the captain, saying, "They're still up there, they couldn't get out."

It was something in the woman's voice that led Danny to believe that this was legitimate. The job was becoming very real. He had only been in the firehouse a few months and couldn't believe the things that he was seeing. Just a few weeks before, he had had to deal with two fatalities at another fire in an electrical substation.

Con Ed Tragedy

It was very early in December, around 1700 hours, right before the change of tours. He was in the kitchen making coffee, putting away the dishes, and getting the meal ready for the night tour. The night tours began at 1630 hours for Danny, who liked to get a jump on his mandatory check of the truck. Other firefighters arrived just before the shift started, but he liked to come in early and relieve one of the senior firefighters.

It was expected that the probie arrive a minimum of an hour before the start of the tour, 0800 for the day shift and 1700 for the night shift. Better to be in an

Figure 3–2. Moving the tower ladder in place to fight a fire in a private dwelling

hour and a half early. This gives the probie time to relieve the senior firefighter and get to checking the tools—saws, Hurst tool, SCBA, portable ladders, and so on.

Suddenly, there was a momentary blackout. The firehouse went dark for about a minute, then all power came back on. It was an eerie moment: everything just went quiet, the TV in the sitting room went blank, the radio in the kitchen just stopped. Almost immediately after the momentary power outage, the tones went off, and Danny knew that whatever this call was for, it had something to do with the blackout.

"Everybody get out, second-due phone alarm for an explosion at the electrical substation."

Captain Dave was working. Tough as nails, never rattled, calm, cool, and collected at all times, he was a larger, powerful man you didn't want to cross. As a firefighter in Brooklyn, he was known as Mr. Bushwick, with a stellar reputation that was all business. He led the company from the front, and if you didn't follow you didn't belong in his company. This instilled confidence in Danny because he felt that he would always be all right if he stuck with Captain Dave.

They boarded the ladder and sped toward the electrical substation. The dispatchers let the companies know that they were getting lots of calls for an explosion at the power plant (figs. 3–3 and 3–4).

Ladder 85 was a seasoned crew, led by Captain Dave, with 25 years on the FDNY working his entire career in the busiest parts of Brooklyn during the War Years. The chauffeur, Fred, was a Vietnam vet who had seen a lot of combat along with Jose, the forcible-entry firefighter and also a Vietnam vet. Fred had been a firefighter in Ladder 85 for 20 years and had also fought all through the War Years in the South Bronx. The other two firefighters, Bobby and Teddy, were both rising stars in the department, being high on the promotion list for lieutenant.

Figures 3–3 and 3–4. Explosion and fire at the Con Ed plant (courtesy of Michael Dick)

Each firefighter on Tower Ladder 85 had a lot of time on the job. Along with Danny on the inside team was Jose. Danny was assigned the extinguisher, as was the norm. He didn't see anything but the can for a few years.

They wound up first into the building. When they got to the landing between the fifth and sixth floors, there was a plant worker lying on the floor with major burns. He was definitely taking his last breaths. Danny stepped over him like he had been doing this for 20 years, unfazed. The captain said, "Let's go, get dat can up here." When they got to the fire floor, the fire was at the end of the hallway to the left. The captain coolly walked over to the source of the fire and ordered him to "put dis guy out."

The substation is an extremely hazardous building, where thousands of volts of electricity pass through from the main power generating station on the other side of the river. Massive cables run under the water, carrying the generated electricity into the numerous tenement and project buildings of the South Bronx. It is so dangerous that it is recommended that the firefighters not carry their metal tools into the area. The problem is not only direct contact with the power grid but also the potential for electricity arcing.

Two firefighters were electrocuted not far from this building in the 1970s at a train yard where the portable ladder that was raised became electrified.

According to the *New York Times*,

> *The victims were carrying out a routine line-switching operation when the blast rocked the seven-story Hell Gate Substation at 132d Street in the Port Morris section at Consolidated Edison Company. The injured man was taken to Jacobi Hospital with third-degree burns over 90% of his body, hospital officials said.*
>
> *He was listed in critical condition in the hospital's burn unit.*
>
> *Two firefighters with minor injuries were also treated at Jacobi and were released.*
>
> *The explosion occurred at about 4:50 P.M. in a narrow sixth-floor room at the substation. . . . The blast involved a switch that linked a transformer to a panel known as a busing bar.*
>
> *The force of the explosion radiated downward . . . but the fire spread rapidly to the seventh floor.*
>
> *Fire officials evacuated 30 to 40 people from the substation, but some Con Edison employees remained at their posts in a fourth-floor control room while firefighters sprayed chemicals on the blaze upstairs.*[1]

---

1. James Barron, "2 Hurt, 1 Fatally, in Con Ed Blast," *New York Times*, December 4, 1986, https://www.nytimes.com/1986/12/04/nyregion/2-hurt-1-fatally-in-con-ed-blast.html.

At the end of the long hallway was the worker who had unfortunately pulled a switch to a bus bar that he shouldn't have. Whatever action he had done caused an explosion, and he had been electrocuted on the spot. Over 10,000 volts of electricity had run through his body. He took the brunt of the electrocution, while his partner was somehow able to crawl 25 ft. to the doorway and make it down half a flight of stairs before passing out.

Undaunted, Danny used the can and put him out. Afterward they continued to search and vent.

Danny acted as cool as a cucumber on the exterior and did not show how nervous he was on the inside. It was part of the job. Working with these men, you wouldn't want to let them see you sweat, and he was not going to let them see him become unglued. When they got back to the firehouse, the guys commented to him on how well he handled his first two "roasts," an old term firefighters used for burn victims. He just shrugged it off.

When he got home the next night, alone in his apartment, that was a whole different story. He was shaken up pretty good, and he couldn't sleep. He kept seeing that vision of the worker on fire. And the smell of human flesh.

He better understood why the firefighters referred to burn victims as roasts, and he couldn't get the image of the victim out of his head. The man was missing the bottoms of his legs and arms, like a roasted chicken. It haunted Danny. He concluded that this was the job, that it was real, and that he better get used to it or it was going to be a rough 20 years, if he could last that long.

Private Dwelling Gets Hot

Danny saw the amount of fire in this private dwelling and knew very well that if anybody was in there it would be a miracle if they were alive. It had only been 3 weeks, and he was dreading having to deal with another roast.

The fire had consumed the first floor of this private dwelling. Danny had a sense of dread because he knew with the amount of fire that this wasn't going to have a good outcome. His memory of the previous fire was fresh, and with his trepidation, he wasn't looking forward to dealing with this again. However, what about his sense of duty?

Flames were everywhere as they entered through the front door. The first floor had two entrances, one in the rear and one in the front, similar to a railroad apartment where the apartment rooms are all in a row like a railroad train. The engine stretched into the front door, but the fire had wrapped around and blown out the interior rear door and up the stairs (fig. 3–5).

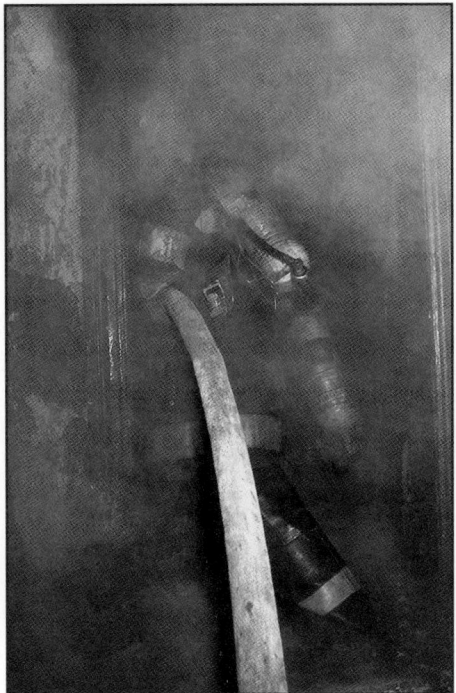

Figure 3–5. Stretching hoseline inside the front door (courtesy of Michael Dick)

They couldn't pass the fire safely to get above using the front stairs, so they needed an alternate route. They grabbed a 24 ft. portable ladder from the tower ladder and put it up to a window above the fire. Danny and Frank Simpson climbed up the front of the building and entered the window, where heavy black smoke pushed out. Any firefighter with any time on the department could have picked up on the fact that the room was about to flash. Being that they were so new, they had no idea what they were getting themselves into.

It was hard to comprehend that anything could be this hot. In those days before Nomex hoods and bunker gear, their only protection was ¾ hip boots and a turnout coat with the collar open at the top.

Danny and Frank stopped and donned their SCBA facepieces, but their ears and neck were exposed, and the crushing heat made their ears feel as if they were being squeezed in a pair of vise grips. Danny tried to pull his collar up but to no avail: the top buckle was red hot, and it was digging into his neck. He felt his ears and neck burning up (fig. 3–6).

If this had happened in modern times, the outcome could have been much different. Sure they wouldn't have been burning up, but they, probably being encapsulated, would have been oblivious to the crushing heat and would have

Figure 3–6. A firefighter's ears and neck burned if they got too close to the fire (courtesy of Michael Dick).

gone farther into the room. The problem is that they would have been so insulated that they would have been caught in the flashover. The fact that they felt the heat gave them warning that they were in a place that they should not have been.

They tried crawling in a few feet, but it was fruitless. They couldn't move. It was like they had entered the pits of hell where the heat was so intense that they could not stand up. Danny and Frank were driven by the heat to their bellies. This situation quickly changed from a search-and-rescue mission to a fight for survival.

They may have been brand new, but they realized that they were in serious trouble, fighting to survive, and that they had to get out quickly.

Ironically, these same two men found themselves in trouble again, having been caught a month earlier in a basement fire that had flashed over. The glow was visible in the rear two rooms of the basement. Brownstone buildings are troublesome when it comes to floor designations. The basement is interchangeably called the first floor. This is more obvious when the front stoop is removed, but when the stoop is intact it becomes confusing. Danny saw that the stoop was intact and let his crew know that there was a good fire in the basement. It happened so fast that they didn't fully understand the gravity of the situation. They were grateful that the two experienced officers were working that tour. Without their experience, the two probies would have been caught in the flashover.

The outside vent firefighter forced a door before the engine company had a hoseline in place. The open door created a perfect flow path for the fire to travel straight for Danny and Frank. It was a flashover situation, and even though they were on their bellies and the heat was unbearable, they didn't respond as quickly as they should have. Captain Albert had to scream at them not once but twice, "Get out!" and then "Get out now!" They weren't going to let the same situation happen again.

Both firefighters quickly realized they were in trouble again, but this time it wasn't the captain who had to tell them. They figured it out on their own. The other flashover had been caused by an erroneous action of a firefighter. He was doing what he thought was right, opening a door to perform a search. Unbeknownst to him, by opening the door, he inadvertently created a perfect flow path, catching them in between the fire and the basement door. This time the flashover, which happens at 1,112°F, was caused by uncontrolled burning. They knew instinctively that they needed to get out of there immediately. Again, they had a covering captain. How Danny wished that his captain was working because he would have figured this out, but he had already gone down the ladder. The covering captain was with them on the second floor near the window and was able to escape.

They crawled back to the window. The captain preceded Frank and Danny onto the portable ladder. The fire was out the tops of the windows at this point and out the whole window just as they were getting onto the ladders.

They bailed out (fig. 3–7), diving onto the 24 ft. portable ladder and narrowly escaping the flames out the window. Danny was able to swing around and make it back down safely. At that point they didn't care what floor they were on or whether or not the ladder was there; they were bailing.

They regrouped on the sidewalk and followed the second hoseline that was stretched to knock down the fire in the hallway. They rejoined the captain and made their way up the burned-out stairway to the second floor behind the engine company (fig. 3–8).

There was still fire at the front of the house on the second floor, but they were able to get into the kitchen and the bathroom. It was then that they found the two victims, one in the bathtub and the other burned up pretty bad on the

Figure 3–7. Firefighter at the window

Figure 3–8. Firefighters climb a stairway with a second hoseline.

kitchen floor. It's ironic that civilians gravitate to the bathroom—they have a feeling that it is a safe place because of maybe the tile or that they can turn on the shower, but either way it doesn't end well. In his career, Danny would always make sure to give the bathroom a good search. He found victims there more than a few times.

Danny and Frank, the two young firefighters, felt defeated. They tried their best to get to that second floor, but there was just too much fire. If it was any solace, the two poor souls had probably succumbed way before they arrived.

If they had continued searching and crawling, surely they never would have made it back to that window. That was the reality, but it still hurt that they couldn't get to the victims in time to save their lives.

4

Clutter, Cocklofts, and Chaos

Scenario: Firefighters find themselves pinned down in a junk-filled living space. A woman living as a hoarder is reported missing.

The building was a wreck. The city ordered the owners to erect scaffolding to protect the pedestrians from the falling brickwork. The numbering and floor designations didn't make any sense. A normal building would have their floors numbered 1 through 6 with letters designating the apartments. In some cases, buildings will have letters to designate the floor, for example, B3 would be on the second floor. The apartment designations here appeared as if someone ran through the building and took random numbers and letters and stuck them on the doors of the apartments.

The building was a disaster waiting to happen. Sometimes you look at a building and think, "What would we do *if* we ever had a fire on the top floor in the rear apartment?"

A Running, Continual Size-Up

Danny had a habit of running scenarios through his mind on his long drive to the firehouse. He felt that firefighting was all about muscle and mental habits. The more you rehearse things in your mind, the more naturally it will play out when it actually happens. He recalled the story of a Vietnam War prisoner at the Hanoi Hilton, who was an avid golfer. Every single day locked up, he played over in his mind each stroke on every hole at his favorite golf course. Upon release after a few years and returning to civilian life, he played the best golf of his life.

Similarly, Danny played out these top-floor apartment fire scenarios in his mind on locating the fire and if it extended into the cockloft. How far has it extended? Getting the roof open and ceilings pulled down were critical to keeping the fire from taking the whole cockloft.

With This Building It Wasn't a Matter of *If* but Rather *When*

When it comes to things going bad at a fire, it's never just one thing. It's usually a series of events happening in a chain. Danny had seen it many times at fires: when one thing goes wrong, it leads to another and another until you are playing catch-up. Firefighters are adept at managing and overcoming one or two obstacles, but when things start unraveling they seem to unravel fast. Water problems are the biggest concerns. Danny knew from experience that when the engine didn't have a positive water source, that was a sure sign of bad things to come.

When a pumper hits a bad hydrant, there is a delay in getting a positive water source. The fire starts to burn unchecked, and before you know it, what starts out as a mattress fire involves the whole room. Once the fire gets out of the room, it extends to the rest of the apartment or house and maybe the floors above . . . and so on. The reality is that these things don't happen in a silo. When things go poorly, the whole operation is affected and the situation gets compounded exponentially.

Where There's Smoke, There's Fire

Just about daybreak one winter morning the company received a phone alarm for a reported fire in a multiple dwelling. While responding, the dispatchers gave an indication that they could be responding to a working fire. Danny could see a black cloud above the rooftop but wasn't sure if it was just the defective oil burner again.

It is the same every year: usually as soon as the weather starts to change and winter starts settling in, the residents of the city start turning on the heating units in the building in which they live. Due to lack of proper maintenance the oil burner nozzle misfires or becomes clogged, causing incomplete combustion and resulting in very black smoke.

It wouldn't be the first time an officer had given a signal for a working fire based on seeing the thick black smoke issuing from the top of a building.

Before Danny got promoted to lieutenant 7 years prior, his lieutenant gave him some very strong advice: "Never give a signal for a working fire until you actually see the flames." (Oil-burner fires can be deceptive. Many an officer has given signals for a working fire when it turned out to be just the oil burner).

Upon arrival, it was very difficult to get a good size-up on the situation. The building was covered by scaffolding (fig. 4–1), and there was no indication of anything going on. That quickly changed. When Danny entered the building lobby it was then apparent that there was a fire in the building. People were streaming down the stairs in their nightclothes. Danny was trying to get a handle on the situation by questioning the escaping tenants. Danny used this technique frequently, trying to get as much intel as possible on the way up. They all gave the same response: there was a fire on the upper floor.

Ladder 96 was first to arrive along with Engine 169. They would be on their own for a while because the second-due units were coming from a distance. The building was a six-story H-type (a non-fireproof, Type 3 mixed construction that has at least two wings connected by a throat) with no well hole, so it was going to be a long, difficult stretch (fig. 4–2). Danny and his forcible-entry team made their way past the panicking building occupants carrying young children

Figure 4-1. Building obscured by scaffolding

Figure 4-2. The stairwell made for a long hoseline stretch.

and clogging the stairs. They were telling the crew that they thought an elderly woman never got out. The outside vent and roof firefighters made their way to the roof with a saw via the adjoining building.

Amid the confusion, Danny lost track of what floor they were on, thinking, "When I get to the fire floor, I'll be okay."

He could not have been more wrong. When he was a firefighter in the squad, one of his lieutenants, Tim McCormack, would always quiz him, "Hey, what floor we on?" (fig. 4–3).

His answer was always the same: "Ughhhh, not sure, Loo."

Danny was so used to working on the floors above when he was in the squad that he lost his ability to do good size-ups. It was always the same in the squad; the chief would send them to go above the fire. He knew that when he got to the fire floor, he just needed to get above. In hindsight he realized that he had felt a little anxious to get to where he was going and failed to take a second to see the bigger picture.

When he finally made it up to the fire floor, the door to the fire apartment had been left open. The smoke in the hallway was banked down to the floor. Danny knew they had to confine the fire and search for the occupants. It was a risky move making entry without the protection of a hoseline, but the occupants

Figure 4-3. During a long hoseline stretch, don't lose track of what floor you are on.

were not accounted for at this time. Armed with just a 2½ gal extinguisher, they decided to push forward. They were met at the front door with blinding smoke and high heat.

The lieutenant and his inside team entered the clutter-filled apartment to search for the reported missing woman. They wound up getting caught with two rooms fully involved and no water with their exit blocked by massive debris. The fire raged in the rear of the apartment.

Danny had a trick he used before the advent of the thermal imaging camera. If you get down low enough, below the smoke, you can see the layout of the apartment. Danny was able to ascertain that there were two options to enter the rear of the apartment. He had seen it many times, a wrap-around island in the kitchen. Essentially there are two ways to get into the rest of the apartment, through the kitchen on the left or straight ahead through the hallway. Either way they were going to have a tough time getting past the shopping carts, boxes, and other clutter stored in that hallway.

Danny and the two other firefighters on his inside team had to squeeze through the narrow opening between the two walls. He knew that if those boxes fell, their way out was going to be blocked and it would delay a speedy exit.

The wraparound hallway was jam-packed with bicycles, plastic storage boxes, garbage bags, and a shopping cart. They chose to go through the hallway because the heat condition was less severe. They made it past the obstacles to the living room without any trouble, but they would be in trouble if they needed to get out in a hurry. While banging their way through the narrow passageway, boxes were knocked over, blocking the egress.

They found the fire in a rear bedroom that extended into the living room. Danny ordered the firefighter with the 2½ gal extinguisher to try to knock down the flames but the attempt was futile. There was too much fire for the extinguisher to make a difference, and the door to the bedroom was burned away. They were pinned down in the junk-filled living room. The only upside was that because of the volume of fire, the visibility was great.

At this point in the operation with everything going on, his situational awareness was zero; he had no idea what floor he was on. The only thing that he knew was that he was in a bad spot with no water. The chief radioed Danny, concerned about the fire spread into the other apartments on the top floor: "Command to Ladder 96, how we doin' on the primary search? Is the fire in the cockloft?" (an extremely important question with respect to top floor fires). Not meaning to be a wise guy, he responded, "Chief, if the fire is on the top floor, then we got fire in the cockloft."

The problem with fires on the top floor is the common void space above the apartments that usually spans up to 3,000 sq. ft. in these types of buildings. The void space can be as much as 2 ft. deep in these types of buildings, similar to a mini lumber yard of dried-out wood.

Danny knew that with this volume of fire in the apartment, it was a good bet that the fire had extended to the void space, but he wasn't exactly sure. Better to err on side of caution.

Danny was getting very concerned; he had been in tight spots before, and this was starting to look like things were going bad very quickly. He started to make his move toward the exit with his team. It was getting very hot, and the second way out was the fire escape in the bedroom that was involved in fire.

The chief knew that the fire was moving fast, and with the delay in water it was time to think about cutting a trench. He ordered the firefighters on the roof to start a trench cut after getting the initial vent hole over the fire (fig. 4–4).

They were now surrounded on two sides by heavy fire that was also in the cockloft over their heads. Danny was worried about not having a hoseline and decided it was time to start making their way out of the apartment. He called for the line but got no response; they needed it immediately. It couldn't get there fast enough (fig. 4–5).

He heard a commotion in the kitchen and was relieved to see the engine team coming into the living room with the hoseline. The officer was not with them, which explained why there was no response. Not only was there no officer with them, but the nozzle firefighter was the new probationary firefighter teamed with another young firefighter. This was, for both of them, their first fire. Danny now had to turn his attention to the hose team.

He could hear the saws on the roof cutting a nice size hole over the fire. As they moved in, he could see the ceiling being pushed down by the roof

Figure 4–4. Trench cut (courtesy of Michael Dick)

Figure 4–5. Firefighter getting hose through a narrow hallway in an apartment building

firefighters. Along with the windows being taken, this gave them a place to push out all the smoke and heat.

In his nervousness and the heat of the moment, the nozzle firefighter opened the bail too early, and the stream from the hoseline knocked Danny's helmet off. Danny knew that he should have been wearing his chin strap, but he often didn't use it. This is why his department was so adamant about using the strap.

Once he had scrambled to get his helmet back on, he reassured the young firefighter, putting his arm around him, coaching him into the rear bedrooms. He instructed the nozzle firefighter to shut the nozzle down and follow him as they knocked down the fire without incident. The two new firefighters did a great job extinguishing the fire in the back two bedrooms.

Now he was able to refocus and complete his ladder company duties. The initial primary search yielded negative results. The secondary was going to be delayed and was done by a different ladder company. It's paramount that a different company perform the secondary search to get a new set of eyes on the area, so Ladder 98 did the secondary, and again the results were negative.

The woman reported missing was not at home. She was visiting her daughter's home in the suburbs.

5

Coordination of Vertical Ventilation with a Fire Attack

Scenario: Two young firefighters failed to maintain their respective situational awareness at the same fire that nearly ended in tragedy.

Get to the Roof No Matter What

"Nothing shall deter the roofman from getting to the roof" was drilled into the heads of new firefighters at New York City Fire Department Academy and reinforced again when they left to be assigned to their firehouse.

Danny was just finishing up his first year of probation in Ladder 85. Kerry was working his first tour in Ladder 97 after transferring up to the Bronx from Downtown. Kerry remembered instructions on reinforcement of getting to the roof when the captain reminded him right away, "Even if people are tossing babies out the windows, you are to continue making your way to the roof."

Assigned to the most senior experienced firefighter, the roof position on the ladder company usually requires firefighters to work alone for some time. That person is the eyes and ears of the incident commander, who relies on reports to make critical decisions that will impact the outcome of the fire.

Vertical ventilation is critical, especially in multistory non-fireproof Type 3 buildings. The products of combustion—if they have no place to go—will bank down and begin mushrooming on the top floor. It took less than a week before Danny found himself responding to his very first fire as the second-due ladder company to a working fire in a five-story multiple dwelling. All new firefighters are assigned to the can (a 2½ gal water extinguisher) and a 6 ft. hook. The newest firefighter stays with the company officer at all times.

The probie makes sure the can—weighing about 40 lb—is filled with 2½ gal of water and then pressurized with 100 psi of air, which forces the water out of the nozzle. The can firefighter places an index finger over the tip of the nozzle, which disperses the water more efficiently.

The 6 ft. hook is a valuable tool in the inside team's arsenal, allowing the firefighter to do a more thorough search by using the reach of the hook to sweep a room. It is also useful to reach out and close a door where heavy fire is located. After the fire is controlled, the probie uses the hook to open up the concealed spaces in the room.

Fire out the Windows

When they pulled up to the front of the building, fire was out three windows on the fourth floor D side of this five-story old law tenement.

It was early autumn, and the days were hot, but once the sun went down, the air was brisk. When the air is cool, the hot smoky air rises much easier because the inside of the building is warmer than the outside; this is known as the stack effect. The older firefighters had this terrific ability to tune out the chaos and focus on the task at hand. As they ascended the stairs in silence, each knowing what was lying ahead, they knew they needed to maintain focus, as opposed to the chief who needs to be able to see the broader picture. They had one job to do here: force entry, search, and vent.

The first ladder company, Tower Ladder 97, was already on scene and had the tower ladder bucket raised, pulling people off the overcrowded fire escapes. Per the department's standard operating procedure (SOP) Ladders 3, they were assigned to the floor above the fire.

Danny grabbed the hook and can off the truck and headed to the floor above the fire, his knees shaking the whole time as he climbed the rickety stairs of this old law tenement. To make the situation more surreal were the shrieks of the women carrying their babies wrapped in blankets as they barreled past them. He thought to himself, "How bad must it be up there if the residents are behaving this way?"

His lieutenant, Stan "The Man" Mercer, was a hardcore seasoned veteran. Stan had been in this area for 25 years and had seen it all. Stan rarely wore his SCBA and had a real ease about him in these tough situations, never rattled.

Danny actually felt comforted by the fact that Stan wasn't wearing his SCBA. He remembered in the academy the day that was the most dreaded, "the smoke house." This was the day that a few of the probies would decide that they weren't cut out for this profession. The instructors brought the probies one squad at a time into the smoke house. You descended one flight of stairs where they had ignited a stack of pallets.

Once all the probies were in position, the instructors ordered the brand-new firefighters to remove their regulators from their mask assembly. Danny did not feel comfortable at all but knew that if he didn't make it through the smoke house, his dreams of being a firefighter were over. The smoke filling his facepiece distressed him to the point where he was going to bail out. He didn't like the acrid smoke burning his throat. He toughed it out and didn't give up even though he was gasping for air.

The fact that Stan was performing his duties with ease made Danny feel that things were okay. This was a great comfort to all the young firefighters he had worked with over the 15 years he had been a lieutenant.

When they got to the floor above, the heat in the hallway was oppressive. Danny donned his SCBA; the smoke was no longer a consideration but now he had to deal with the heat.

Another aspect of the smoke house that Danny wasn't crazy about was the radiant heat. He had been naïve in a sense at the academy; never being close enough to an actual fire, and watching a lifetime of movies where actors ran through flames, he had never expected a fire to be this hot. As they descended the smoke house where a few pallets were burning, he was taken aback by the amount of heat that was being generated by this pile of wood. Any skin that was exposed was burning. It was a new experience for him.

This situation was worse. Again like the smoke house, he was on top of the fire. All that heat was coming through the floors like he was standing in a frying pan. They weren't able to stand, but at the same time the floor was burning through his rubber boots. He managed to be able to duck walk, where he wasn't quite standing erect but not crawling either.

Danny then felt the force of the hoseline hitting the ceilings below him and the heat beginning to subside a bit. The oppressive heat was now dissipating as a result of the water cooling the area and the heat being pushed ahead of the hoseline. As soon as the roof bulkhead was opened, it completed the ventilation process.

This made Danny very nervous, and he started to question his decision to become a New York City firefighter. He had made up his mind right then and there, in a flash, that if God allowed him to survive this he would resign when they got back to the firehouse. He knew that he had made a big mistake. He couldn't understand how his lieutenant was so calm and cool. He didn't even have a mask on!

On the flip side, if Stan could have seen through the smoke, he'd have seen a face full of terror. The few training fires and the smoke house at the fire academy were nothing like this. The lieutenant calmly asked him, "Hey kid, didja feel dat?" referring to the sudden change in conditions on the floor above the fire.

As the lieutenant uttered those magical words, Danny felt as though God had heard his pleas because it was all of a sudden tolerable in the hallway. He no longer felt the impending doom and was actually able to stand up. The roof firefighter had popped the bulkhead, and the engine was now moving into the fire apartment on the floor below (fig. 5–1).

Danny

Captain Dave was acting chief again, so Tower Ladder 85 had a covering captain filling in for the night. Having a covering captain is no different from when you were in high school and they sent in a substitute teacher for the day. That captain is "just there," as they say in the department, "for the night." Besides having the covering captain, they also had a firefighter from a neighboring engine company detailed. Danny was still a probie, so he was assigned to carry the can.

Figure 5–1. Interior of bulkhead

The older, more experienced engine firefighter had the forcible-entry position with the irons. The inside team, known as the forcible-entry team, is made up of the officer, the can, and the irons.

Danny was concerned because he lacked experience and was on a team with two firefighters he didn't know. He was getting used to working with Dave, who had become more like a father figure. Danny's father had left him when he was 3 years old, so he never really knew what it was like to have a father. He loved watching shows like *Father Knows Best* and *Leave It to Beaver* in the 1960s, and for that half hour he would escape into that world, but after the show was over it was back to the reality of living in a tenement with his mom and little sister.

The firehouse was his home away from home; he had never experienced such intimacy with men before. He lived with his grandmother since high school and had no real male role models to look up to. His own father came back into his life during his second year of college, but that relationship never could go back to such where his father could become a mentor.

Danny was impressed with the values and family life that most of the firefighters in the firehouse exhibited. Danny emulated some of the guys there and eventually got married and moved to a house up in the "country," upstate New York.

Even though Danny was still new to the job, he knew the importance of the team. After being there close to a year now, he became comfortable working with the firefighters on his shift. But the three firefighters there that night had never worked together before. They were in reality three separate individuals.

It's like putting three musicians together and expecting them to make great music: they may know the notes that are expected but the song won't sound the same as a trio who plays together all the time. On the other side of that coin, the outside team—the chauffeur, outside vent, and roof firefighters—that night were a well-oiled machine, senior firefighters who had worked together at numerous fires. In the fire service it is all about the team. They all know the SOPs of their department, but do they really know each other?

Kerry

This was his first night tour working in Ladder 97. Kerry had just transferred to this busy South Bronx ladder company from a ladder company in the downtown part of the city. It was a busy hot July night, and Ladder Company 97 had two probies working that night on the inside team with a very experienced lieutenant, Bobby Boyle. Kerry was assigned as the roof firefighter. In Kerry's former company the buildings were very different, primarily loft buildings, whereas his new company's district was loaded with old law tenement buildings that had been built at the end of the 19th century (fig. 5–2).

Kerry was surprised that the officer would assign him to the roof position on his very first night tour in his new company. He was familiar with the department's SOP for roof access, in priority order:

1. The adjoining building
2. The aerial/tower ladder
3. Rear fire escape

Unless there was some variation where a building may be isolated or next to an adjoining building of a different height, firefighters in the South Bronx stuck to the policy. There are very few "never" and "always" rules in the FDNY, but one is "The firefighter assigned to the roof position is to never take the interior stairs to access the roof bulkhead." Another peculiarity about Kerry's former district was that most of the buildings had front fire escapes that had a gooseneck ladder that went to the roof. Here none of the fire escapes that had street access had gooseneck ladders (fig. 5–3).

Being it was his very first night in the South Bronx put Kerry at a great disadvantage. There was no way that Kerry could have understood the different types of buildings in the area in one night. The area he came from could not have been any different, as it was loaded with high-rise office buildings and commercial loft buildings.

Before 1968, buildings constructed in New York City that were not Type 1, fire-resistive, construction were required to have a second means of egress. In

Figure 5–2. A row of tenements

Figure 5–3. Gooseneck ladder, so called because, like the drop ladder, it is a vertical ladder that is attached to the top floor landing and extends to the roof, where it is curved and bolted to the roof parapet. It looks like the neck of a goose (courtesy of Tim Klett).

Type 3 tenement buildings this usually meant a metal fire escape attached to the front or rear of the building.

The metal fire escapes have a drop ladder that is held to the structure by a metal hook on the second floor. It is the duty of the outside vent firefighter to use a 6 ft. hook to push up on the drop ladder and release the ladder from the hook.

Danny

At exactly midnight, the house watch teleprinter jumped to life, sounding like one of those old stock market ticker tape machines. The teleprinter was an antiquated system that printed out a response ticket for the fire alarm on a sheet of paper. The paper was rolled and placed on top of the machine. It was fed through some rollers and, when the alarm came in, the paper dropped down and the teletype went across the paper and printed all the information for the incident. It made a unique sound as it printed the address of the reported fire building, "tatatatatatat."

It continued until every letter of the response was printed. Some of the fire-fighters were so dialed in that as soon as they heard the first tick, they were already turning the companies out.

Tower Ladder 85 was assigned first-due ladder to a report of a building fire a few blocks away from the firehouse. Danny was at the house watch, and he turned out the companies by blasting the old navy foghorn three times. The house watch area is the communications hub of the firehouse. One firefighter is required to staff the desk at all times, responsible for answering the telephone, recording all movements of both companies in the company journal, dealing with any civilians that may come to the firehouse, and acknowledging and turning out the companies for all alarms.

"Wham! Wham! Wham!" the foghorn blared.

Danny then shouted into the pitch box that had speakers throughout the firehouse, "Everybody goes! First-due phone alarm, fire."

The firehouse burst to life, the house lights were all turned on, and the run was acknowledged via the teleprinter. Not much has really changed these days except that the teleprinter has been replaced by a computer screen that has the ability to store all the incidents. The teleprinter/computer's only purpose is to provide a mechanism to produce a response ticket.

The system has evolved throughout the years in three distinct phases: (1) The bell system where the firefighter on the house watch had to literally count the bells then refer to the "running board" that had all the companies' box assignments posted. (2) The voice alarm system used the PA system that was connected to the firehouse from dispatch. The dispatcher would announce the alarm details over the voice alarm in the firehouse. This system is still in place and tested every morning at 0845 to be used as a backup. (3) The teleprinter/computer prints out a response ticket. Any subsequent information is then broadcast over the department radios inside the apparatus.

"Engine, ladder, and battalion 10-4 send."

Danny left his boots and turnout coat next to the house watch because he didn't ever want to be the last one dressed. While getting dressed he could hear the dispatchers on the house watch scanner radio calling the chief to let him know that they were getting a few calls on this one. He opened the apparatus doors and went to his spot on the truck behind the officer on the passenger side of the apparatus.

The engine company had a 1970s-era American LaFrance 1,000 gpm pumper that had a distinctive whine. The chauffeur revved it up like a stock car racer waiting at the gate for the starter to give the green light and then started to do a slow roll out of quarters while the firefighters got on the rig. Once they were all on board, the engine bolted out of quarters like a rocket.

The chief waited patiently by the doors, as he always did, ready to close them, but the ladder company wasn't moving. The chief was anxious and was hearing the reports from the dispatcher on car radio and knew that this was going to be a good fire. He had waited long enough and decided that he needed to go and took off. The chief as a courtesy closes the apparatus doors; he is supposed to let the companies out first, generally the engine followed by the ladder. Rare exceptions where the truck is assigned first due would have the truck leave before engine. It was apparent that the ladder company was having a serious delay, so he closed the engine door, got into his car, and rode with his aide to the fire.

The firefighters from Tower Ladder 85 were saddled up with their adrenaline pumping, and the radio was full of chatter. The engine on scene gave a signal for a working fire. The covering captain still wasn't on the rig. In the past there have been times that a company may go down a firefighter and respond, but this was different—they needed the captain. The senior chauffeur was extremely aggravated. He yelled to Danny, "Hey, kid, go upstairs and see what the hell is going on."

Danny raced upstairs to the office and busted in. The captain was in his dorm with the TV blaring and the AC on at full blast. He shook the captain, "Hey cap, we gotta run."

The delay had set them back and put them behind the eight ball, which meant that they were now second due. Normally companies will not make any attempts to "beat each other in," but this was a no brainer; Ladder 97 beat them in by a few minutes. Tower Ladder 85 had accepted the fact that they were not going to arrive before Ladder 97, even though this building was dead center between the two firehouses. Because the South Bronx was so densely populated at the turn of the 20th century, the companies were relatively close to each other. Legend has it that the distances were dictated by how far the horses could travel without exhaustion, but either way they were close to each other.

The delay felt like an eternity. In reality it was at most 2 minutes, but it was enough of a gap that 85 had to assume the second truck responsibilities.

Ladder 97, arriving on scene well ahead of Tower Ladder 85, had no choice but to assume the duties of the first-due ladder. Engine 116 was on scene and stretching a hoseline and was going to need assistance with forcible entry and ventilation.

The ominous glow from the fire was visible when 85 turned the corner onto the avenue. There was so much fire out the windows on the upper floor that it looked like they had a vacant building fire, but this was an occupied building. Because they were second-due ladder, their assignment was to go to the floor above. When they arrived on the fourth floor, the fire floor, they encountered heavy fire in this railroad apartment on the right side of the old law tenement.

Railroad apartments (aka railroad flats) are so named because of their resemblance to a railroad train where all rooms are in a line. Two apartments per floor mirror each other with two entrances plus a fire escape. The norm is that the kitchen and bathrooms are in the rear with one entrance and a living room followed by a small bedroom with a front entrance to the last two bedrooms. There is an air and light shaft on both sides of the building (fig. 5–4).

The apartment was burning front to rear: every room was involved, and there was so much fire that the smoke was seeping through every orifice in the apartment. In a fire like this, the second-due position is the more precarious one because the first-due truck waits for the engine to get water and then follows them into the apartment. The second-due ladder company, on the other hand, needs to go above this inferno. One "always" or a very strong recommendation is that in an old law tenement, firefighters must ensure that there is a charged hoseline in place before going to the apartment above. The engine was in place with a charged hoseline.

The fire was right behind the door in the front. In these older tenements, the transoms over the door that used to provide ventilation were only covered over with a thin piece of plywood, which was the only thing separating them from the raging fire as they made their way past to the fifth floor.

Because of the crushing heat in the hallway on the fifth floor, they could tell the roof was not open yet and something was amiss. The first-due lieutenant from Ladder 97 was screaming to his roof firefighter, "Ladder 97 to roof, we need the bulkhead opened; we're getting hammered!"

Figure 5–4. Outside a typical railroad flat

Kerry

Kerry wasn't sure of the best access to the roof. He saw that his company's aerial ladder was being used in the front to remove civilians, so that wasn't an option.

Billy Strava, the outside vent firefighter from Ladder 97, could see that Kerry was having problems. Kerry was staring at the front fire escape, confused (fig. 5–5). He was thinking that since the tower ladder was being used, he'd take the fire escape. Billy saw this and stopped him in his tracks before he committed and directed him to the building next door with the entrance around the corner.

Kerry was feeling the pressure. It was his first night in his new company and he didn't want to start off on the wrong foot. Fire was out every window in the front of the building, so he knew it was even more critical that he vented the roof. He knew all too well what it felt like when the roof was not vented. He imagined the scene inside the building on the fire floor and the floor above where the heat and smoke had nowhere to go. He was sure the firefighters inside were getting slammed.

The seconds felt like minutes as he trudged his way up the stairway of the exposure on the B side. Kerry was moving slowly because he wore his SCBA, which firefighters assigned to the roof normally didn't wear back then. Firefighters

Figure 5–5. Front fire escape

in those days wore leather boots and a shorter coat with no SCBA. They liked to get to the roof quickly and be mobile.

When he arrived on the fourth floor of the B exposure, the smoke was already banked down to the floor because the window in the public hallway had been broken. He was forced to don his SCBA facepiece and then he continued on his way to the roof.

Danny

Conditions in the fire building had deteriorated horribly, Tower Ladder 85's inside team donned their SCBA and made their way into the apartment on the floor above to get away from the heat conditions in the public hallway and found it no better inside. The fire had possession of the front two rooms to the right that were off the street. Danny went left toward the rear of the apartment to search and vent; he was confident that if things got bad, he would just pop out onto the rear fire escape. When he made it to the rear windows to vent, it was then that he realized his almost grave error: there was no rear fire escape (fig. 5–6).

This building was one of the rare exceptions, a railroad flat with only a front fire escape. He was in big trouble, and he went from being confident to a near panic. The way he came in was his only way out and now it wasn't an option. The fire burned through the floor in the living room, cutting off his egress. He

Figure 5–6. Rear of tenement with no fire escape

said a prayer, asking God to get him out of this jam. Due to his inexperience, he had never considered that the railroad apartment that went from the front to the rear had a fire escape in the front but not the rear.

Back in the turn of the century it was required to have a second means of egress. Sometimes the builders, to save money, attached them to the front. Danny was totally taken off guard. Normally the roof firefighter would communicate this with the chief over the radio for all to hear: "Ladder 85 roof to command, be advised that there is no rear fire escape on this building."

Radio transmissions are not just for the chief but for everyone on scene. Something this critical should have been transmitted. Danny got angry because his captain would have known that and slowed Danny down. Thankfully the engine appeared at the door with the hoseline, allowing him to make his way to the front door.

Kerry

By now Kerry had made it to the roof of exposure B (fig. 5–7). The smoke was billowing out of the air and light shaft between the fire building and exposure B (fig. 5–8). He was disorientated and made a right turn when he exited the bulk-head, making his way to the back of the building instead of the front. Being that the bulkhead faced the avenue instead of the rear, he was heading in the wrong direction. Corner buildings are the most dangerous types of buildings firefighters face because they don't conform to the norm. At this point his lieutenant was

Figure 5–7. Exposure B roof building

Figure 5–8. Air and light shaft

pleading with him to get the roof open. Kerry responded, "Ladder 97 roof 10-4 almost there."

Being on the roof of a burning building, especially if the fire is on the top floor, is one of the most dangerous parts of the job. When the smoke is this bad, you need to just stand still until the wind shifts or use a hook to feel your way around. When Kerry realized he was going the wrong way, he felt the pressure even greater now because he knew that he had failed at his task, to get the roof vented. He needed to move forward, even though he wasn't comfortable with the whole situation. The firefighters on the fire floor and the floor above were counting on him. Kerry lost his situational awareness and started to make bad decisions:

- His lieutenant was pressuring him to get to the bulkhead and open it.
- There was a tremendous amount of fire and smoke venting out of the shafts.
- He was unfamiliar with these types of buildings.
- He was disorientated.

By the time he managed to make it to the parapet wall that divided the two buildings, he was frazzled, having lost his focus. He did like he was trained by using his 6 ft. Halligan hook to feel for a deck. The tool landed on solid wood, indicating that it was safe to traverse the parapet (fig. 5–9). What he didn't realize was that there was a very narrow gap that was not visible because of the heavy fire and smoke condition. When he stepped off the adjoining building, he dropped down the shaft.

Figure 5–9. Parapet

Timmy

While this was happening, the second-due roof firefighter Timmy Galvin made his way to the roof via the D side. He knew that he was now second due so he decided take a different route to ensure one of them made it up there. It is not uncommon to find a fence separating the buildings, which is why some firefighters prefer to take the aerial ladder and forego the adjoining building.

When he got to the roof he saw a set of tools on the roof deck, yet the bulkhead door was still shut. "Where is 97's roof firefighter," he said, confused. "His tools are near the shaft and door is shut? Oh God, I hope he didn't . . ."

He popped the bulkhead door and then went to investigate the shafts and to see if he could find 97's roof firefighter. Fire blowing out the windows into the shaft made it impossible to see anything. He yelled down and got no response.

Kerry

Kerry couldn't believe this was happening. He was dropping into this shaft like a rock at an incredible speed past the fire, bouncing from window to window. The only thing he could think of was to pray, his life flashing through his mind as he begged God to forgive him for all his past wrongdoings. When he hit the ground, the same SCBA that slowed down his descent drove into his back.

Kerry heard someone shouting down from the roof but was in so much pain he couldn't respond due to his massive injuries, blood filling up in his lungs. He tried to answer the voice but was only able to eke out a meager response on his Handie-Talkie: "Mayday, help me. It's Kerry, I fell down the shaft."

The captain from the second-due engine heard it and transmitted the Mayday.

Conditions were not improving inside. Without a radio, Danny heard a Mayday transmitted through his captain's radio and his first thought was that someone fell through the hole on the fifth floor. Engine 114 was with them on the fifth floor and was moving the line toward the front of the building. He pictured someone from Engine 114 being unaware of the area that had burned through.

Rapid intervention teams (RITs) didn't exist then, so when firefighters were in trouble, other firefighters on scene would have to divert from the mission to address the downed firefighter. This night was no different, so when the Mayday was given, the focus changed from fighting the fire to getting to the downed firefighter.

Fallen Firefighter Is Rescued

Kerry was extremely fortunate that he fell onto a ton of rubbish in the shaft. The nozzle operator from the first-due engine was the first to get to him. He shimmied his way down a drainpipe from the fourth floor.

Kerry had fallen approximately 70 ft. into an enclosed shaft. Because of the excessive amount of rubbish in the shaft, the two basement doors were covered over. The only way to gain access into the shaft was to breach a wall (fig. 5–10). Some of the firefighters made their way to the outside wall of the enclosed shaft, while the rest of the crew continued attacking the fire. The chief transmitted a third alarm to compensate for the resources committed to the rescue. In essence, the chief was dealing with two distinct incidents, or an incident within an incident. The situation here was dire. Kerry was bleeding out internally and his liver was held together by a thread, nearly split in half.

While the fire was still raging above, Kerry was lying on a pile of debris, surrounded by four walls. The rescue company went to work with the Tower Ladder 85 chauffeur on breaching the wall. They used pneumatic drills and mauls to bang away on the two courses of brick that were in their way.

After working tirelessly for what seemed like hours, they managed to make an opening large enough for a firefighter to squeeze through. While the firefighters and EMTs worked on Kerry, they continued to enlarge the hole. Kerry had lost a lot of blood and was in extremely critical condition but was alive. They put him on a backboard and removed him to the awaiting ambulance.

Figure 5–10. Firefighters breaching wall

Incident Review and Reflection

The next morning, Danny and the rest of the crew from Tower Ladder 85 returned to the building to do an after-action review. On the roof, Kerry had fallen incredibly into the narrowest part of the shaft, barely a foot wide. On one side they could see the marks in the brick where the SCBA dug into the brick-work, a straight line, and on the other side they could see the marks his fingers made on the wall.

Danny caught up with Kerry later on to discuss what had happened. They found it was easier to talk about their shared experiences as they were true to their Catholic faith. Danny shared with him his stupidity in going to the rear of the building and finding his second egress cut off.

Kerry had nothing but praise for his rescuers. The nozzle firefighter from Engine 116 climbed his way down a drainpipe and was the first to get to him. Fellow firefighters placed him on a backboard and brought him to the ambu-lance. He shared with Danny a spiritual experience. While he thought he was in his final moments of his life, he asked God for forgiveness and then his Aunt Grace, a nun in Ireland, appeared to him. While he was fighting for his life, she was by his side.

He miraculously healed and returned to active duty.

6

Communicating an Emergency in NYC

> **Scenario:** Engine 116 was tasked at two separate fires where they had to make multiple rescues and extinguish fire at the same time.

New York City installed Emergency Reporting System (ERS) boxes to replace the old pull boxes in most neighborhoods. Civilians used to pull the lever on the alarm box—a pull box—sending a signal to the main dispatch office. In the 1970s, the newer ERS boxes had two buttons, one blue to summon the police and one red to summon the FDNY (fig. 6–1).

The idea was to cut down on high false-alarm rates. The ERS box allowed the caller an opportunity to talk to the dispatcher. The FDNY did not dispatch a unit unless someone *spoke* into the alarm box between 0800 and 2300 hours. "No contact, no response" was the policy in the FDNY. If it was after 2300 hours, they would send one engine.

Engine-only operations in the big city sometimes happen when there are multiple alarms dispatched simultaneously. The engines ran all night from 2300 hours until 0800 the next morning. Neighborhood kids pushed the ERS box at all hours of the night. It was difficult to avoid complacency because it was expected that any alarm from an ERS box would be a false alarm. Once in a while an ERS box would come in, the engine would go out, and a minute later the tones would go off again. Whenever the tones would go off *after* the engine went out, it was almost always a working fire.

The ticket usually read "FILL OUT ALARM 2ND SOURCE," meaning two additional engines and two ladders with the battalion chief.

Engine Company 116 shared quarters with Danny's tower ladder during the War Years and did close to 10,000 runs, and those are only the ones that they recorded. Danny was fascinated about the amount of fire alarms they responded

Figure 6–1. Emergency Reporting System box

to back before he got to the firehouse. The ladder company was so busy that they had two sections, Tower Ladder 85 and Ladder 85-2.

Firefighters got beat up from 1500 hours (3 p.m.), when school let out, until around 1800, going to a few working fires every tour. Danny asked Big Tom one day: "How did you stay mentally prepared for every single alarm?"

Tom explained to him about the "three rings" when the dispatchers received a phone alarm for a fire, "They would call the firehouse and give us three rings. When we got the call, we knew we were going to a fire." They didn't have a teleprinter, so the house watch firefighter had a pen and paper handy to record the particulars of the alarm. If the dispatchers felt that the call had a sense of legitimacy, they called the firehouse house watch desk and gave the phone three quick rings. This was a tip off to the firefighters that someone or the dispatchers were getting a few calls on the box. This is similar to the concept of getting an ERS box followed up by a few phone calls.

The three rings were a way for the dispatchers to break the companies out of complacency, letting them know that they were getting a phone call in addition to the pull box. In those days it was so easy to get complacent with companies going on 10,000 runs.

Fire in the Fish Market
with People Trapped

Still on probation, Danny was working a mutual swap in the engine when a senior guy needed off. He volunteered to work the shift, just running around on another hot summer night in the ghetto. The streets were vibrant, filled with people drinking and listening to their boom boxes at all hours. Typically, the streets were alive with people hanging out on the corner drinking beer, listening to rap music on their boom boxes, or playing dominoes in front of the corner grocery store. Kids in the schoolyards played pick-up basketball. The summer night in the South Bronx was no different from a crowded midtown Manhattan city street scene.

Around 0200 hours they received an ERS, no contact. The house watch firefighter called it out sounding a little annoyed: "ERS, get out engine." Danny hadn't been around long enough to be complacent. He was just thrilled to get on the rig. The rest of the crew was half asleep, tired from the busy night of running. There were no other calls to support this box call. The engine turned the corner, and they could see a massive column of smoke in the distance, a working fire in a four-story multiple dwelling, a mixed occupancy Type 3 corner building. Type 3 construction was ordinary brick and wood, and the crew needed to know these facts because it would dictate the SOP. A Type 3 suggests that the fire on the first floor in the store will communicate fire, heat, and heavy smoke to the floor above the fire. Danny sometimes wondered how they could get 10 calls for a pot burning on the stove and then one call on flames shooting out five windows.

The ground floor housed a fish market that spanned the entire building; the other three floors were residential, one apartment per floor. The A–D corner came to a point like a slice of apple pie, making for a tough size-up. These buildings don't conform to the expected norms (fig. 6–2).

Fire was ripping through the fish market where numerous panicked people were showing at the windows on the second floor above the fire (fig. 6–3). With this much fire and people trapped, the captain of the engine made a great decision to split the team and address the life hazard while extinguishing the fire. In these situations with limited manpower, there are only three options:

- Fight the fire and attempt rescues
- Perform rescue only
- Fight fire and shelter in place

Danny was assigned to a ladder company, so the captain ordered him to get a portable ladder and start removing the occupants. The engine only has one type

Figure 6–2. A corner building that does not conform to expected standards

Figure 6–3. Fire out the windows above a commercial property

of ladder on the apparatus, a 24 ft. extension ladder. Full of adrenaline, Danny was able to pull off the ladder by himself and put it up to the window. In his haste and nervousness, he set up the ladder upside down. He climbed the ladder and removed about five people, but one woman panicked and refused to get on. Danny tried to physically pick her up and carry her down the ladder to no avail. After what seemed like an hour he eventually coerced her onto the ladder.

He reentered the window and was driven to his belly. It was lights out with extremely high heat and smoke conditions. He was right on top of a store fire that was fully involved, and he knew there were more people trapped inside. It felt like he was in a frying pan. He had two choices, go straight or go right, and he decided to do a right-hand wall search toward the bedrooms in the front. Danny heard a brief moan but couldn't orient himself. He continued on the path he chose toward the front bedrooms.

He found another woman, but by this time the ladder companies were on scene, so he passed her off to one of the firefighters from his ladder company. As he made his way toward the front of the apartment, he found a woman in the bedroom. She was hysterical. Danny started guiding her toward the window where he bumped into a firefighter from Tower Ladder 85, whom he asked to take her out so he could continue his search in the last bedroom. He was hugging the front wall, on his way toward the front of the building. He had no tools to help with his search so he fanned out using his hands. He heard a transmission that they had found two more victims. The second-due ladder had come up the interior stairs and found two people right behind the front door, unconscious.

After the fire, when he was alone, he beat himself up for being so dumb. He really thought that the bedrooms were the right move. He made the decision to go toward the bedrooms and it turned out that the people were behind the front door to the apartment. If he had to do it all over again, he still would have gone for the bedrooms. In reality he couldn't do it all; there were too many people trapped and those bedrooms needed to be searched. Had he gone left, he wouldn't have found the woman but would have found the two kids. Either way, all survived. The problem with corner buildings is that each layout is different.

A decade later Danny reflected on the Fish Store Fire to drive home to himself the importance of doing a diligent search. When he was lieutenant in a ladder company, they responded second due to a report of a fire in a brownstone building. The dispatchers received a report that a woman was trapped on the top floor. To get to the box alarm location required a climb up a massive hill. He remembered that the rig just couldn't get there fast enough. At one point when they got to the base of the hill, the chauffeur stood up from his seat to get the rig to move up the hill. When they arrived, Danny, the lieutenant, jumped off the truck and ran toward the building. The civilians were telling him in the street, "Forget it, she jumped."

He didn't want to hear it; he needed to get to the top floor. He continued running, bolting up the stairs taking them two at a time. He didn't wait for his forcible-entry team. When he got to the landing on the top floor, he just kicked in the door. He didn't want to believe it, but there was no denying the truth: the window was smashed, broken glass on the bed, and a bloody handprint on the upper part of the wall. He looked out the window and down in the rear yard was this beautiful young girl, broken.

The tragedy was that the fire was just a small rubbish fire at the base of the staircase. The smoke may have been thick but the fire was small. She must have opened the door and saw the smoke and panicked. He still can't get that image out of his mind.

———

Firehouses are not always notified of fires by the ERS or dispatch. Sometimes, they get a "fire knock" or verbal alarm from a civilian. For this reason it is important to always have a firefighter at the house watch area.

One quiet summer day, the engine company remained at quarters while the ladder company was out procuring the meal. The peaceful morning was broken up by a frantic banging at the door. Danny's buddy Billy would call it the fire knock as he opened the door and three very excited teenagers pointed to the building across the street.

Fire was blowing out the windows on the thirteenth floor. The house watch firefighter turned out the company and notified the department dispatcher that they had a working fire across the street.

The building was a 19-story high-rise fireproof multiple dwelling. The officer and three crew members threw on their coats, boots, and helmet and raced across the street. They received a report from the dispatcher that a woman was still in the apartment. The chauffeur pulled the engine across the street and hooked up to the hydrant that happened to be directly in front of the building opposite the fire department connection (fig. 6–4).

So far this was going like clockwork. Each of them grabbed the 2½ in. hose packs from the side of the engine and the lieutenant Jimmy Wendler grabbed the standpipe kit. The kit contained everything they needed to operate from the standpipe outlet:

- Door chock
- 18 in. pipe wrench
- Outlet in line pressure gauge
- Adaptors in case the threads are different
- Wheel for outlet in case it was missing

Figure 6–4. Engine in front of building hooked up to hydrant

- $^{11}/_{16}$ in. controlling nozzle
- Spanners

Jimmy ordered Danny to grab the irons—a Halligan tool and axe—as well as his 2½ in. rolled-up hose (fig. 6–5).

A building occupant had verified a report of the woman trapped, so his adrenaline was raging. They were short staffed but had everything they needed. Ladder 85 was going to be seriously delayed, so it was going to be just them for a while.

They called for the elevator and waited. They knew that they had a big fire on their hands, and Danny felt like a soldier aboard a military LTV, waiting to go into battle. The seconds seemed like minutes waiting for the elevator, and the longer they waited the greater the dread. Danny wanted to get up there and face this demon. After what seemed like 10 minutes, one of the kids from the building told the firefighters, "Don't waste your time; the cops took the elevator already and the other one's broke."

This was going to be rough: climbing 13 flights with over 100 lb of equipment with ¾ hip boots was an arduous task. Danny was in great shape but Larry and Artie were not. Larry was a heavy smoker, and Artie was at least 50 lb heavier than when he left the fire academy 5 years earlier. They began their ascent, but Artie was already winded by the sixth floor and they still had a fire to put out.

Figure 6–5. A 2½ in. rolled-up hose on side of engine

When they got to about the ninth floor, they heard some banging, metal on metal: "Bam! . . . Bam! . . . Bam! . . . Bam! . . ."

Jimmy asked the three exhausted firefighters, "What the hell is that racket?"

The banging got louder as they ascended the stairs. By the time they got to the 12th floor the banging had stopped. Jimmy went to the fire floor with Danny while Artie and Larry hooked up to the standpipe outlet on the 12th floor in the "A" stair, which contained two fireproof enclosed stairways. There were two stairs that crisscrossed on alternate floors (aka scissor stairs). Years later the FDNY would institute a policy whereby one stair would be used solely for attack and the other would be deemed the "evacuation stair," the idea being that civilians could escape from the floors above in a smoke-free environment while the other stairway would be used to attack the fire. The caveat is that the door to the fire floor must remain closed at all times, lest the stairway become contaminated and negate the smoke-free environment.

They found the two police officers dragging an unconscious woman who had been directly behind the front door out of the apartment. They were extremely lucky that the wind was not blowing in because, with that amount of fire, the wind could have easily blown the fire toward them. The two officers had forced the door using a tire ramp.

Danny headed back down to the 12th floor; he was winded but remarkably was still ready to get in and put this fire out. They flaked the hose out using the floor below, onto the stairs, and brought the hose directly to the fire apartment. Jimmy saw that the hose was properly flaked out and called down to Larry to start water in the 2½ in. hoseline. Danny assumed his position behind the burly nozzle firefighter, and the big guy took that 2½ in. hose in like it was a garden hose. Artie was powerful and practically dragged Danny with him.

Artie was gasping for air inside his mask because he was down to a quarter tank, but his company pride wouldn't allow him to let Danny have the nozzle. After he extinguished one room, he had no choice: his vibrate alarm was going off (his cylinder had 30 minutes of air) and he couldn't take another step. He reluctantly passed the nozzle to Danny. He followed the hoseline directly to the stairwell door and collapsed from complete exhaustion on the floor of the stairwell landing.

When it was over, the lieutenant was fuming. The two police officers quickly scurried out of there, nowhere to be found. He later discovered what had happened. The two police officers were standing on a post next to the firehouse when they saw the flames out the windows. Instead of alerting the firefighters, who were only feet away inside the firehouse, they decided to run into the building.

They took the elevator to the fire floor and kept it there by placing something to prevent it from closing. The husband met the police officers in the hallway and informed them that his wife was still trapped inside. They located a mechanics ramp from a neighbor and started to bang the door down. The wife was found directly behind the door.

The bottom line is the whole building always needs to be searched, every hallway on every floor and all stairways. The thick black smoke will penetrate every floor, and it is impossible to know what people will do in moments of extreme panic. This is even true for trained professionals.

7

Multilevel Floor Collapse: Withdrawing Firefighters I

> **Scenario:** A firefighter's gut feeling prevents firefighters in the fire apartment from heading down to the basement from the fourth floor when the floors give way.

Danny was happy to be working with his favorite crew: Lt. O'Brien (FOB), Billy, Pauly, Victor (recently transferred to Tower Ladder [TL] 85 from Engine 116), and Freddy, "The Barge." Danny was assigned the outside vent (OV) position. He rarely if ever was assigned to the OV position; he usually worked with Big Tom, who would never let Danny take the OV position. Billy, who loved the water can, was assigned that position. Victor was on the forcible entry and Pauly had the roof position.

At this time TL 85 was still responding to lots of vacant building fires, and tonight would be no different. The firefighters finished up the evening meal when the tones went off: "Everybody goes. Engine second due, truck first due, report of a fire in a vacant building." The address was for a four-story tenement building, which was one of three tenements connected in a row. The crew had previously been there several times in the past few weeks, where apparently someone was trying to burn these down (fig. 7–1).

The firefighters from TL 85 knew these particular buildings: even though they were connected in the front, each building was separated from the others with firewalls. There were air and light shafts between the buildings as well (fig. 7–2).

A predictable cycle of events followed with these buildings. They started with owner's neglect, fires started, and residents moving out, making the building vacant (fig. 7–3). This provided the chance for vagrants, drug dealers, and squatters to take over. More fires happened.

Figure 7–1. Remnants of a vacant building fire

Figure 7–2. Interior shafts

Figure 7–3. Vacant structures can be havens for homeless or transient people where fires can happen.

When TL 85 turned the corner onto the boulevard, the familiar glow shone off in the distance. FOB banged on the glass that separated him and The Barge from the crew in the rear. "They got it goin' good dis time."

Engine 114 was first to arrive. Anytime TL 85 went north, 114, which was a single engine company, would usually get there first. "Engine 114 to dispatch, we got a workin' fire in a four-story vacant multiple dwelling."

Engine 114 took the hydrant on the corner and stretched to the front door, but they needed to wait for the ladder company to force the blocked-up front entrance before they could get the hoseline to where it was needed on the second floor. The entrance was, as usual, covered with 8 in. cinder blocks (fig. 7–4). The Barge positioned the tower ladder in front of the building in case the chief decided to go to a defensive attack, where they would set up a master stream and extinguish the fire from the outside.

For the time being, Engine 114 was preparing to initiate an offensive attack. The firefighters got geared up while TL 85 worked on removing the cinder blocks covering the front door. They saw the fire out the windows and knew this was taking out the whole floor. They made sure to button up their coats and pull up their boots. Even with their boots pulled up, the hot scalding water and embers still somehow found their way down the narrow opening above the knee below the bottom of their coat.

Figure 7–4. Cinder blocks on front entrance (courtesy of Michael Dick)

The fire was on the second floor, D side of the 50 ft. × 75 ft. Type 3 vacant multiple dwelling. The whole apartment was going from the front to the rear. Billy and Victor went to work on breaking the cinder blocks with an 8 lb maul. In addition to the cinder blocks, the front of the building was boarded up with Department of Housing and Urban Development (HUD) windows, pieces of plywood attached to both the inside and outside of the window with threaded bolts held in place by 2 × 4s (fig. 7–5). The city tried their best to seal up these vacant buildings, but usually any attempts to secure the building were easily thwarted by the locals who wanted to use the building for their own needs.

Danny and The Barge set up the tower ladder, which took longer compared to setting up an aerial ladder. With the tower ladder you couldn't take any short-cuts because the massive boom was extremely heavy and it was important that all four jacks were down as well as both tormentors. The rule of thumb was using a 6 ft. hook to judge the distance the tormentors needed to be fully deployed. When the tormentors were down the tower ladder resembled the wings of a small plane when looking down from the bucket.

The time lost, in Danny's opinion, was well worth the extra effort because he was able to fly the bucket to multiple windows with ease. He was responsible for VEIS (vent, enter, isolate, and search), which meant that he had to remove these HUD windows before he could get inside to perform a primary search. Conducting a search in this vacant building was exceptionally dangerous, the reason being that you could never be sure if the occupants had cut a hole in the floor or if there was some partial collapse. For safety reasons it is important to use a tool, whether a Halligan or a hook, to probe the floor ahead while hugging the wall.

Figure 7–5. Department of Housing and Urban Development windows

Although the building was vacant, TL 85 treated it as occupied because such buildings usually were full of transient people who mostly used them as drug dens. Danny set up the bucket and operated in the front of the building (fig. 7–6).

Before Danny could get to work on the HUD windows, he had to first bring the roof firefighter up to the roof and drop him off. Because all three buildings were vacant, the bucket was the safest and surest way of gaining access to the roof.

Billy and Victor were slow in forcing the cinder blocks covering the front door. The arduous work of removing the cinder blocks meant there was going to be a delay in getting water on the fire. For each minute they were delayed, the fire grew. Danny dropped Pauly off at the roof and dropped down to the second floor to get to work on the HUD windows. The trick was to smash the 2 × 4 cross piece with the point of the Halligan tool and loosen the bolt that held the two pieces of wood together. Once the 2 × 4 was loose, it was easy to remove the wood covering.

Some windows had a metal covering and the best tool to use was an axe (fig. 7–7). Even though power saws replaced a lot of what the axe was used for, it was still the best tool for this job. Danny used the back of the axe to smash the middle of the metal, causing the outer edges to pop out and exposing the nails. Once the nails were exposed, he used the Halligan tool to open the window.

When the fire took off, the chief became concerned. Although he preferred to fight the fire from the interior, he knew he may have to use the master stream. The upside of attacking the fire from the interior was that it was more efficient

Figure 7–6. Bucket operating in front of the building working on Department of Housing and Urban Development windows

Figure 7–7. Firefighter using a hook to break through metal window coverings

and there was a better chance of not giving up the whole building. The downside was that firefighters were more exposed and it would be a long, drawn-out operation. He liked to get units back in service quickly and not tie them up with a vacant building fire. He weighed the risks and decided that the building was still in pretty good shape.

Danny was able to get back to the window off the front fire escape after taking the other three from the bucket. The fire had complete possession of the second floor on the D side and had extended a bit to the third floor through the voids in the bathroom. The hoseline was ready to start the fire attack. Danny entered the apartment on the B side to search where he could; he knew that there were people living here but found no one.

His nozzle operator, Bobby, was a large barrel-chested guy who moved the 1¾ in. hose like it was filled with air, and he made quick work of the fire. Billy and Danny opened up the ceilings to check for any extension. The second engine, 116, stretched another hoseline to the third floor to deal with the little bit that got up through the pipe chases in the bathroom. Engine 114 finished mopping up the fire on the second floor. The decision to perform an interior attack paid off: instead of spending hours pouring water on the building and eventually losing it, they spent about an hour. The building would still be salvageable for future renovations.

Rekindle was unavoidable in vacant buildings, unless you wanted to spend 5 hours trying to open up every wall and ceiling. Sometimes it was good enough, and if you had a hot spot, so be it. This night was no different, and a few hours later they got called back for smoke coming from the same vacant building. The building was wide open now so there was no problem gaining access to the fire area.

There was a decent smoke condition out the windows on the top floor. Since there had been no prior issues with the integrity of the building, the chief thought it best to just go up and handle it from the interior. The firefighters made their way up to the top floor and found a bedroom going pretty good in the rear apartment.

Billy attempted to knock it down, but the fire was too much and they were going to need a hoseline. FOB called down to Engine 114 to stretch a 1¾ in. hoseline to the top floor. They were already stretching because they had assumed that a line was going to be needed. The captain had no problem stretching; if they didn't need the line they could repack it, no big deal.

Engine 114 had the hoseline up there in no time. FOB directed them to the rear bedroom where the inside team and Danny were. Billy had a bad feeling. He wasn't comfortable with something, and his gut instinct was telling him that they shouldn't be there. Maybe he felt the building shake or something, but he just knew something was wrong. He grabbed Danny by the back of the coat and pulled him out of the room, saying, "Sparky, this doesn't feel right." As soon as he pulled him, the whole D side collapsed in what is known as a pancake collapse. This is where a floor gives way and collapses on top of another floor all the way to the bottom, similar to a stack of pancakes.

Larry, the nozzle operator from 114, disappeared into the belly of the building and left Charlie, the backup firefighter, teetering on the edge of the precipice. This wasn't the first time Billy had pulled Danny from harm; he seemed to have a sixth sense when it came to these situations.

The entire building was in danger of collapsing, and the situation remained unstable. After Larry disappeared, Danny, Billy, and Victor headed straight for the basement. They knew exactly how to get to there because they had just been in the building a few hours prior. Billy and Danny followed Victor, who led them to an opening in the sidewalk that looked like a rabbit hole that led straight into the basement.

They were expecting the worst, as there was no way Larry could have survived the collapse. The fourth floor had given way when the weight of the firefighters and charged hoseline entered the bedroom. It had collapsed onto the third and then all the way to the basement.

The basement was a pile of rubble, with lathe and plaster, 2 × 4s, broken 2 × 8 in. joists, and pipes teetering over their heads. They picked a spot that seemed like a logical place to start looking. For all they knew, Larry could have been buried under tons of debris. The three firefighters worked feverously to free him. Just as Billy's intuition had helped them avoid disaster, his same gut instinct helped them find the exact spot where Larry was buried.

They began to dig, and miraculously the form of Larry's body appeared. He looked like a baby curled up in the womb of his mother, but not moving. Danny called out to him, "We're comin' to get ya brother."

He didn't expect a response, but there was a groan.

Victor then said, "When this is over and you're outta there, we'll get a beer."

Unbelievably, the downed nozzle firefighter replied, "I don't drink!"

Filled with dread, Danny had figured they were retrieving a body, not rescuing a brother firefighter. But they were all relieved. He was talking!

The chief called out to Lt. O'Brien, "85 move aside, let the rescue company get in there."

The Barge shot the chief a look. He didn't get that nickname for no good reason; it was highly unlikely that he was going to let anyone past him. TL 85 had found Larry and they were going to take him out. The chief stepped back and didn't say another word.

The three firefighters were close to getting him out when one said, "We got 'em chief; we'll be out in a minute."

Danny was positioned at Larry's back. Billy and Victor were on either side. The firefighters on the first floor passed down a small backboard and got it over to Danny, who straddled over Larry to reach down and cut the straps on his SCBA to slide the board down behind his back. The three of them pulled him out and carried him over to the rest of the firefighters waiting on the first floor. They passed him up to the awaiting rescue company, who took him out to the waiting ambulance.

Miraculously, Larry suffered no major injuries and spent a few days in the hospital. He was cleared and did a few months on restricted duty, eventually returning to full firefighting duty.

Multilevel Floor Collapse: Withdrawing Firefighters II

Scenario: A brand-new firefighter saved the lives of at least seven firefighters.

A week before Christmas, firefighters from Engine 116 (E 116) and Tower Ladder 85 (TL 85) responded to a fire in the New York City Housing Authority (NYCHA) housing projects near the firehouse. A tenant had failed to water a live Christmas tree in her apartment. The building was a low-rise fireproof multiple dwelling that didn't have a standpipe, which meant that E 116 was going to have to hand stretch a 1¾ in. hoseline. These stretches could be a bit difficult and time consuming because of the outward-opening windows and the small landing over the front doorway entrance.

Engine 116 needed to bring a Clorox bottle (fig. 8–1) with 75 ft. of ¼ in. diameter rope to use from a window on the floor below to speed up the process (fig. 8–2). TL 85 entered the apartment with the 2½ gal extinguisher to try to confine the fire. They figured there'd be a delay in getting the 1¾ in. hoseline in place.

Fire completely filled the living room. Billy got enough stream on the fire from the 2½ gal extinguisher to darken it down, but they were going to need that line. He used his index finger to break up the stream, making the can more efficient. After expelling the can on the fire, he and Danny made their way to the back bedrooms to do a primary search. The hoseline came up very quickly and knocked down the rest of the fire, which resulted in a routine project job (fig. 8–3).

Because the building was of Type 1 fire-resistive construction, the firefighters never gave much credence to fires in these structures: the contents were generally light and the fire would not normally extend beyond the original fire apartment (fig. 8–4). Such a fire in today's buildings would be much different and

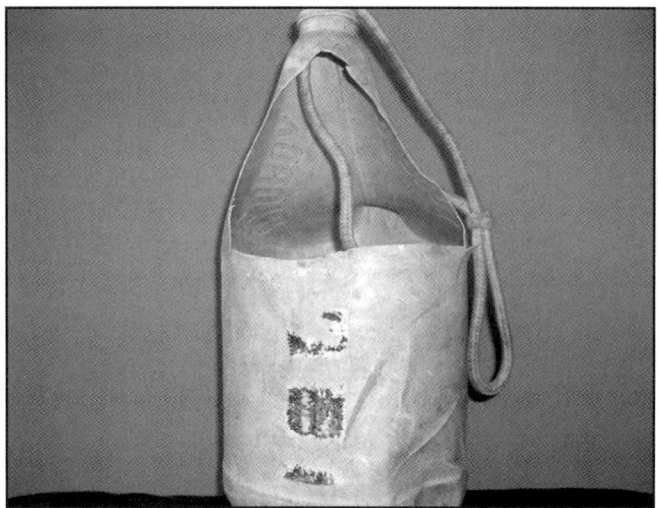

Figure 8–1. Bottle of Clorox

Figure 8–2. Rope stretched some 75 ft. from the top window of the building

Figure 8–3. A typical layout in an apartment following a fire

Figure 8–4. The aftermath of a fire in a low-rise fireproof multiple dwelling

create a whole new dynamic with changes in products of combustion from cellulose-based products to petroleum based. This, coupled with the potential for a ventilation-limited fire coming in contact with an air source, has made these fires much deadlier.

Devastating Fire to the Family

It may not seem like a big deal, but to that woman and her three little kids, the apartment and its contents were her world. She lost everything, including all the gifts that were under the tree. It is heartbreaking to see a poor family with all their possessions in ruins. When they got back to the firehouse, Billy rallied the firefighters together to try to do something to give those kids a proper Christmas. They collected enough money to buy the kids some Christmas gifts, clothes, diapers, and other essentials. They planned to show up at her apartment on Christmas Eve with everything they collected to surprise her.

Manpower was short that Christmas Eve and someone needed to work overtime. Danny volunteered because he had no children and his wife would be spending Christmas Eve with her parents. If you couldn't be home, being at the firehouse was the next best thing. It was tradition in the 16th Division on Christmas Eve that Engine 117 and Ladder 88 would host a Catholic Mass. Companies that were close enough were invited to participate on Ladder 88's apparatus floor.

Chief McNulty and his driver Bob Byrne combined had about 70 years in the FDNY, so it was a safe bet they were off for Christmas Eve. This meant that someone from either E 116 or TL 85 had to drive the acting battalion chief. This vacancy caused a shortage in TL 85, so a detail firefighter was coming from midtown. Since Danny was working on overtime, he volunteered to drive the chief. He really wanted to be in his own company that night to be there for the presentation to the family, but they couldn't have a detail from outside the battalion let alone the house drive the acting battalion chief.

When the detail showed up at the firehouse, his gear looked like it had just been bought from the fire department supply store; everything was shiny new. His uniform was crisp. He was from Engine 56 in midtown but lived close to the firehouse, so he was happy to be working there that night. The guys in TL 85 had arranged for him to get "early relief" so that he could be home when his kids got up to open gifts that Santa brought.

The detail firefighter, Darryl, had just got out of the fire academy, and when he arrived, Danny went over to the battalion to drive. He had never driven the chief before, so he had no idea what to do. When he relieved Bobby, Bobby assured him that all he had to do that night was drive the chief.

The firehouse was buzzing; on a normal night, the meal was a big deal in the firehouse, but tonight even more so. The turkey from the local *vivero* (Spanish for "nursery," it actually was a live farm that normally held chickens, but near the holidays they brought in turkeys) had been cooking all day. The guys were wrapping gifts and getting ready to buy some last-minute items for the family.

Captain Livolsa, who was called in to be the acting battalion chief, wanted to stop by his firehouse before Mass, so they headed out, leaving the festivities. His firehouse was in full holiday mode as well: guys cooking up a storm, Christmas music blaring, one guy dressed like Santa. They stayed for a bit, had some food and laughs, but the time was getting close for the Mass. They said their good-byes and headed out.

On the way to the Mass, they stumbled across TL 85 operating at a water leak in a six-story multiple dwelling on the boulevard. They turned the corner onto the street and saw a woman frantically jumping up and down at the end of the block. His gut told him, without even seeing the condition on the wide two-way avenue, that something bad was happening. He called the guys from 85, saying, "Forget that water leak, I think we got a job."

When they got to the corner, the woman pointed to the middle of the block. Fire was blowing out a storefront into the avenue past the sidewalk (fig. 8–5). The store was typical for the South Bronx, an occupied store on the first floor with the rest of the five stories unoccupied. Captain Livolsa radioed the dispatcher: "Battalion 55 to dispatch, we have a working fire at this location. Transmit the all-hands on arrival, send an additional engine and ladder."

Danny was lost for a moment but smart enough to move the chief's car out of the way to leave the front of the building open for TL 85 and E 116, which converged on the building at the same time from different directions. They were all business: 116 hit the hydrant in front of the building while 85 set up the tower ladder to cover the front of the building. The fire was impinging on TL 85 because both windows had failed. Danny grabbed the Halligan tool from the engine and made his way to the front door of the store.

The deputy chief arrived and ordered Captain Livolsa to go to the floor above the store and supervise operations there. The fire had already extended to both apartments on the second floor and was racing up the rear of the building. The second-due engine, 114, arrived on the second floor with a 1¾ in. hoseline (fig. 8–6).

This was a problem because 114 was overwhelmed, trying to deal with fire in two apartments with only one hoseline. The captain was directing the nozzle firefighter to alternately hit the fire on both sides of the hallway. With only one

Figure 8–5. Fire blowing out into the street (courtesy of Michael Dick)

Figure 8–6. Hoseline operating inside vacant building

hoseline it was difficult to get a handle on the fire, which was growing rapidly to the floors above. The only safe place was the public hallway.

The inside team of TL 85 was working with the engine in the store on the first floor. Bobby Norton was a super athlete; in the Bronx at times he would be mistaken for New York Yankees outfielder Dave Winfield. As quick and agile as he was, by the time he made his way up to the roof via the adjoining building, the fire had already got through the roof (fig. 8–7). He attempted to ventilate the roof, but conditions got worse quickly. He crawled back into the adjoining building where he radioed his officer: "Ladder 85 roof to Ladder 85, the fire has complete possession of the whole rear of the building."

Billy took the bucket up to the fourth floor because making entry into the second or third floor wasn't feasible. There was a possibility that some people may still be inside the building. He teamed up with Charlie Blackmore from Rescue 30. They entered the apartment from the front window where they were immediately forced to the floor, as the heat was immense. As soon as they entered they were already fighting for survival because conditions changed so rapidly. They needed to get out of there quick. No search would be performed. They dove out the window for the bucket of the tower ladder, whether it was there or not.

Captain Livolsa remained on the second floor. He said to Danny, "Hey kid, do me a fava and check out what's happenin' in da store and get back up here."

Figure 8–7. Fire through the roof

The fire was out of control. Too much fire and too little resources. E 116 was making some progress with their 2½ in. hoseline, but the fire was already in complete possession of all floors in the rear of the building. These railroad flats have the bathrooms and kitchens in the rear of the apartment, which contains many voids where they run the plumbing and vent pipes. The fire had gotten into the void spaces and took off, and between that and the probable copious amounts of diesel and gasoline that had been poured through the building, the fire was well advanced even before the units got on scene.

Arsonists would use both fuels in their mischief. The gasoline, which is much more flammable because of its lower flashpoint of −40°F, would be used to start the fires, which would burn rapidly. They would then use diesel, which has a much higher flashpoint of between 125°F and 180°F, to finish the job.

When Danny made it to the store, he noticed Darryl, the detail firefighter, standing inside near the front door with his can and 6 ft. hook. E 116 was in the rear of the store and had extinguished most of the fire with their 2½ in. line. There were still a few hot spots here and there, but the fire in the store was knocked down.

Fire continued raging on the floors above. The incident commander was getting ready to pull everyone out and go to a defensive operation. E 116 was finishing up, hitting the remaining pockets of fire. At the front of the store Darryl was screaming that there was fire, and the seven firefighters in the rear seemed at first to ignore him, but Darryl wouldn't stop screaming.

Lieutenant Fagan had finally heard enough and ordered Howie to bring the line to the front of the building. They all vacated the spot in the rear and went with the line. When they got to the front of the store they discovered it was just a few embers burning. No sooner had they vacated the back of the store than there was a loud crash, sounding like a freight train. The whole rear of the building collapsed in the exact spot they were standing just a minute prior. Four stories of debris swiftly pancaked down (fig. 8–8).

The nozzle firefighter, Charlie Reagan from Engine 114, was now hanging upside down, entangled in BX cables and lath. Lieutenant Fagan and Ted Kelty from E 116 quickly got to him and freed him. Reagan miraculously survived the collapse, coming out unscathed, despite tons of debris in a pile. He was placed on a backboard and taken to the hospital.

The Christmas gifts were delivered while the battalion was out at the acting battalion chief's firehouse. When the firefighters from E 116 and TL 85 showed up at the door, the woman broke down in tears.

Figure 8–8. The aftermath of a pancake building collapse

9

Fighting Fire in a High-Rise

Scenario: An officer and two firefighters are caught on the wrong side of a high-rise fireproof multiple dwelling fire when a firefighter on the roof vents the windows without consulting with the officer from the ladder company.

Rarely did FDNY receive only *one* call for any type of incident in a high-rise fireproof multiple dwelling (HRFPMD) building. Calls to the 19-story NYCHA buildings (affectionately called "the projects") kept firefighters of the 55th Battalion plenty busy (fig. 9–1).

Figure 9–1. Typical high-rise fireproof multiple dwelling

On any given day they responded to water leaks, gas leaks, stuck elevators, burnt food on the stove, garbage compactor chute fires, rubbish fires in the hallways/stairways, and the occasional apartment fire. Even the occasional apartment fire wasn't really thought of as a big deal back then. If asked the next day on how the tour went, most firefighters replied, "We had a few runs, not much going on last night, just a project job."

Fires in these buildings have unique SOPs. If the fire is above the seventh floor, firefighters must rely on the erratic, sometimes nonfunctioning firefighter service elevators to get them to the upper floors of the building. It was not unusual in these types of buildings for the firefighter service to be out of service. This was very disconcerting for the firefighters (fig. 9–2).

The firefighter service function requires a key so firefighters can recall the elevator to the lobby and manually operate it. The first objective in a HRFPMD is to locate and verify the fire floor. If there is a report of a fire on the tenth floor, firefighters will take the elevator to the eighth floor and walk the rest of the way. There is never a reason for taking the elevator to the reported floor, which is a disaster waiting to happen. Firefighters never want to be in a position where they may wind up in an elevator that lands on the fire floor.

When the elevators are in firefighter service, they remain in the control of the fire department. If the fire was located on the seventh floor or below, firefighters could easily walk up the stairs. Some firefighters climbed stairs if the fire was a reasonable distance and depending on the circumstances.

Figure 9–2. An elevator out of service is not welcome sight to firefighters.

Fighting Against Complacency

Responding over and over to the same building wore down these firefighters. Like the boy who cried wolf, they guarded against becoming desensitized and complacent. Unlike the private dwelling where one or two families may reside in a two-story structure, these structures are 19 stories with eight apartments per floor. Anytime there is any sort of emergency, whether a gas leak, stuck elevator, or fire, the dispatcher receives numerous calls. Most of the time, nothing will be showing when firefighters arrive. It is impossible to do a proper size-up from the street. The incident commander is totally reliant on the intel they get from the dispatchers and from the reports from the firefighters inside the building. On occasion there may be flames or smoke visible from the street (fig. 9–3).

Conditions can change very rapidly in a fire of any kind in a HRFPMD. A fire in a kitchen (like with food on the stove), if given the right conditions with water, can erupt into a fireball in an instant. If the windows fail with a strong wind blowing into the apartment, a small amount of fire can morph into

Figure 9–3. A high-rise fireproof multiple dwelling with flames out the window

blowtorch conditions. This is why it is critical that firefighters follow their SOPs in these types of buildings. Even if there is little or no wind on the ground floor, this doesn't reflect the conditions that may be encountered on the upper floors (fig. 9–4).

Danny had been in the firehouse only a few months and was still not comfortable with his heavy equipment. He considered himself to be in good shape; he worked out in the gym, ran almost every day, and did pushups every night. Still, this job was kicking his butt. Most, if not every, tour he carried the 2½ gal extinguisher along with the 6 ft. hook. His SCBA and the rest of his gear felt like he was dragging around an extra hundred pounds. His helmet was a new-style leather helmet that was top heavy and didn't fit right. The rubber boots were the biggest problem. They seemed damn near impossible to move around in, let alone climb stairs (fig. 9–5).

It was early in the tour. Tower Ladder 85 received a phone alarm for a report of fire in the 19-story HRFPMD building next door. As usual, Danny was on the house watch and turned out the companies: "Everybody goes, first due, smoke, top floor."

Because the report was for the 19th floor, most of the residents didn't smell it, and the rest of the occupants on the top floor were probably at work. When they arrived on location, there was no indication of anything happening from the street level. In a HRFPMD building such as this, it would be expected to receive lots of calls any time there is a fire, especially a fire on a lower floor.

Figure 9–4. Wind blowing in from an upper floor

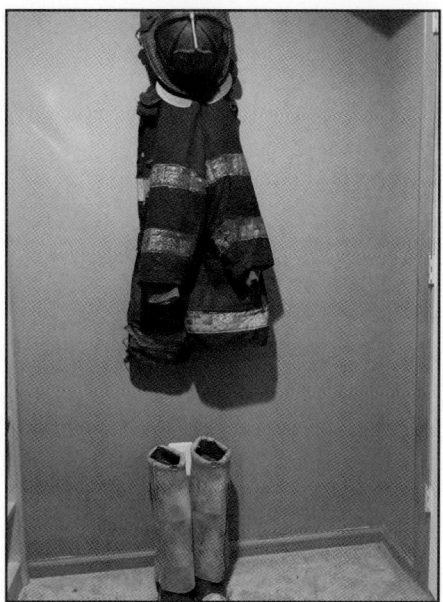

Figure 9–5. Firefighter's full gear circa 1986

It was difficult to determine the location of the fires based solely on the calls. Later on in Danny's career when he had more experience, he learned to discount the calls for smoke, but as soon as the dispatcher mentioned a fire in the apartment, he would focus on that floor. If the report was "Dispatch to Engine 1, we are now receiving a report for a fire in apartment 7J," he would climb with his company up to the seventh floor.

They arrived at the lobby of the building to again find that the elevators were out of service. They were going to have to walk up the 19 flights. This was no problem for the irons firefighter, Phil.

They didn't like the cumbersome rubber boots, especially in the high-rise buildings. The six-foot-three Phil was a natural and excelled at everything he did, whether it was playing basketball or forcing doors. However, for Danny this was going to be a tough climb (fig. 9–6). He was struggling to keep up with Superman. In those days, the senior guys like Phil wore shorter coats and leather boots.

It wouldn't have surprised Danny if Phil could have jumped up to the 19th floor in a single bound. Although he was 10 years younger, he struggled to keep up with him. It wasn't the weight of the equipment or being out of shape; it was the damn boots and helmet. When they got to the 10th floor, the irons firefighter quietly took the water can off his back and continued without missing a beat. When they reached the 19th floor, the water can waited for him at the top of the stairs. He grabbed it, and they made their way to the fire apartment.

Figure 9–6. Interior stairs of high-rise fireproof multiple dwelling

After the door was forced, they found themselves in the small hallway outside the fully involved living room (fig. 9–7). The bulk of the fire was the large couch that was up against the windows.

Old Lt. Murphy with his distinct accent gruffly yelled to Danny, "Knock down da fire wit da water can." This was Danny's favorite part of the job, so he donned his SCBA and crawled into the living room with his 2½ gal extinguisher. Early on in his career, he took advantage of plenty of opportunities to use his tool.

He got briefly mesmerized by the beauty and raw energy of flames consuming the room. For a split second, he thought it'd be a shame to disrupt this serene moment by turning his weapon upon the enemy to beat it down into submission.

Danny knocked down the fire so Phil and Lt. Murphy could search the back bedrooms. He had learned from the best and became effective at using the water can to its maximum potential. Billy taught him to place his right index finger over the nozzle to disperse the water more efficiently. Billy liked to keep the water can a "little light," believing the extra pressure (100+ psi) compensated for the little less water.

Figure 9–7. Outside the apartment in the hallway

Danny got the water from the water can onto the couch and knocked down a good amount of the fire, but not all of it. Lt. Murphy knew that once the water can was expelled there would be no more water until the engine got there.

Even though the fire was still burning a little, he decided to go past the living room to complete the primary search. Because they had to force the door, he couldn't be sure that the apartment was empty. They made their way past the fire toward the rear. The engine was delayed even more so getting up to the fire floor because they had to carry that hose up to the 18th floor. The engine company was still in the process of hooking up the hose on the floor below.

As they headed to the back bedrooms, they heard the sound of glass breaking. Vito, the roof firefighter, banged out the windows in the living room without conferring with Lt. Murphy. The smoldering couch became an inferno as the wind blew into the apartment toward the open front door. Phil grabbed Danny into a side bedroom and closed the door while the lieutenant ran back toward the front door. They were now trapped on the other side of the raging fire with no second means of egress. The fire raged outside the door. If the engine didn't get there quick, it would be time to start thinking about getting out the window.

If Phil was concerned, he certainly wasn't showing it. He acted as cool as a cucumber. Vito and the other firefighters on the roof prepared to deploy the lifesaving rope (figs. 9–8 and 9–9). Ironically, the same guy who got them into the jam was now going to have to save them.

There were no personal safety systems in those days. Some firefighters took it upon themselves to outfit themselves with their own personal escape system, but these guys didn't have one.

The fire was starting to burn the top of the door and smoke was pouring in through the cracks, but they were still okay for the moment. Phil's calmness during the ordeal had a profound effect on Danny, who thought, "Hey, if he's not worried then I guess we're okay." The fact that he kept cool and didn't get rattled made a lasting impression on Danny that stayed with him throughout his entire career.

It didn't change the fact that Phil was pissed off at Vito. He told Danny, "He took the damn windows; he never checked with the lieutenant to see if the hose was in place."

Not a moment too soon, Danny could tell that things were going to be all right. He saw Phil's reaction when the engine got water in the hoseline and began extinguishing the fire. They heard that magical sound of 250 gal of water hitting the walls and ceiling. Danny opened the door a crack to where he could see Engine 116 moving in. Dean, the nozzle firefighter, looked almost ghostly through the flames, quieting the raging inferno.

Figure 9–8. Lifesaving rope is traditionally carried by the FDNY "roofman" and is 150 ft. of nylon with a hook at each end.

Figure 9–9. Roof of a high-rise fireproof multiple dwelling

10

You Don't Need Much Water If You Get It Where It Needs to Go

Scenario: The new lieutenant's pleas for booster water went unheeded in this single-family private dwelling.

The two new lieutenants were happy to finally get a temporary assignment in Engine 179. It was a slower pace than they would have liked, but it was a place to call home for a while. They were assigned in opposite groups, so they would often relieve each other and spend at least an hour after the tour kibitzing. This temporary assignment was a blessing for them both.

The pair were full of life and helped set up drills with the firefighters every tour they worked. For the guys who didn't care to drill as much, they were out of luck because they were going to drill if either of the two was working. This particular night Danny had a split crew, two fairly new firefighters and two who were more senior. One of the senior firefighters, "Binky," had worked in the same battalion as Danny. He had transferred up to the slower place a few years back when his engine was disbanded. The other firefighter had over 30 years as a senior chauffeur.

As usual, it was a quiet night, with a few false alarms and medical runs. Danny held a drill before the night meal in their full personal protective equipment (PPE). His motto was "Train like we play." The two younger firefighters were assigned the nozzle and backup. The lights were turned off in the basement and they put their Nomex hoods over their facepieces.

He set the basement up to replicate a basement of a private dwelling, which was all they had in their district. When they got to the top of the stairs, he instructed them to "get down there quick . . . if ya hang at the top of the stairs, ya gonna fry."

They developed a system where they kept the line moving and would dash down the stairs. Danny could see the progress after every drill. His only lament was that they didn't do it for real enough. It was not like when he was a fire-fighter, when they would work almost every night. He reminded his crew every time they drilled that "just because you're in a slower company doesn't mean you can't be as good as the busy companies."

Never Enter the Fire Area
Without Water

Engine 179 received a phone alarm near midnight for a fire in a private dwelling. They responded to an affluent, remote part of the city, which meant they would be operating by themselves for a while. The quiet tree-lined street had no indication of anything happening. Danny caught a faint whiff of something burning, and he had a feeling they had a working fire; it just wasn't apparent at the moment (fig. 10–1).

Figure 10–1. Fire on second floor of a private dwelling

The building was a two-story flat roof private dwelling. He threw his mask on his back and headed toward the house, while the rest of his crew stayed by the back step in preparation to stretch a 1¾ in. hoseline. He forced the front door pretty easily with his officer's tool. After 9 years on the job, he had gotten pretty good at forcible entry.

It was one of the things he missed about being a firefighter. In his department, officers supervised firefighters and didn't get involved in the hands-on work. It was one of the transformations from firefighter to officer that hadn't fully sunk in yet. This was an exception because there was no ladder company in yet to force the door. He could see the glow at the top of the stairs (fig. 10–2).

He ordered his crew to stretch the hoseline to the front door, knowing that the department had a strict policy to "never enter the fire area without water." An entire private dwelling is considered part of the fire area.

Climbing the stairs, he eventually found a fire contained to the bedroom only. Danny ordered the chauffeur, "Engine 179 to chauffeur, give us water in the line as soon as it's at the front door" (fig. 10–3). He made a quick dash past the bedroom off to the left of the stairs to check out the rest of the second floor. Luckily

Figure 10–2. Fire at the top of the stairs

Figure 10–3. Firefighters waiting at front door with a charged hoseline

there was no one home, so he returned to the stairs and headed down to get his crew. He thought that this was going to be a great experience for his young nozzle team.

Danny and the young firefighters were getting anxious. He couldn't imagine what could be causing any delays. They had recently inspected the hydrants on the street in this neighborhood and apparently no one was vandalizing them in this part of the city. If you looked like you didn't belong in the area, the residents would let you know in no uncertain terms that you weren't welcome, a far cry from any neighborhood watch where Danny worked as a firefighter.

Danny called again with a little more forcefulness in his voice, "Engine 179 to chauffeur, gimme water," and then, after a brief silence, he got this response: "I'm not on a hydrant."

Danny replied, "That's okay, just gimme a booster," which was again met with silence. The lieutenant returned up the stairs to see what was happening, thinking, "Maybe I can get the door closed; if only we could get some water in that line, we could knock the fire out."

But it was now out into the hallway and there was no chance of confining the fire. He returned to the front door to get the hose team and still no water was in the hose. Again, he ordered the chauffeur to send water into the dry hoseline: "Engine 179 to chauffeur, gimme booster water, now!" More silence.

He grabbed Binky, the firefighter he knew, and told him to go and see about the delay. The fire was now out the bedroom windows in the rear and heading toward the front of the house, making the top floor untenable. He now had no choice but to stay at the middle of the stairway and watch helplessly as the fire consumed the top floor, cursing himself for not bringing a water can, which if he had had earlier, he could have at least used to knock the fire down.

After what seemed an eternity, there was water in the hoseline. Ladder 215 finally showed up at this point, but there was nothing for them to do until they got the fire knocked down. Chris was a probie at his first fire. Danny could tell that he was nervous, so he put a hand on his shoulder and calmed him: "We're going to the top of stairs, and you are going to hit the fire, and it will be all good."

"10-4 lieutenant" was his reply. He was nervous but remained calm. He trusted Danny, and he knew that he had lots of experience, which was a comfort to him.

The chief who arrived on scene changed everything instantly. He was a screamer, and Danny knew from past experience that he became easily unglued, even on the fire scene. This seemingly one-room fire was starting to unravel.

The first issue was the delay in water, and he had seen over the course of his career that any time there were water issues, bigger problems always followed. In this instance, once the water problem was resolved, they were ready to move forward. He was reminded of the Swiss cheese theory: when holes in the cheese lined up, disaster resulted.

The chief screamed over the radio that the fire was out the windows in the front and that there was fire in the cockloft. Danny grabbed Chris, the nervous probie, and looked at him straight in the eyes through his mask facepiece, reassuring him that everything would be okay.

They proceeded up the stairs with the nozzle open, knocking down all the visible fire. They moved left and finished the rest of the second floor. Ladder 215 was able to get into the room after the fire was knocked down to check for extension, opening walls and pulling ceilings. There was no fire in the cockloft.

Danny felt that the fire was adequately extinguished, and he didn't like to overdo it with the water. Anything that was smoldering could be tossed out the window; this was somebody's home, and he didn't want to do any more damage than was already done.

He had empathy for fire victims. This scene reminded of his own life, when fire destroyed everything. As a young boy, his family lost what little they had when their small apartment burned out. He could still remember the smell from that fire some 30 years later.

He radioed the deputy chief who was on scene: "All fire knocked down; we can shut down the line."

"10-4, nice job 179. Come on down with your crew."

Ladder 215 was just finishing up overhauling. The lieutenant called command, "Ladder 215 to command, urgent we have extension; we need a line."

Danny couldn't believe what he was hearing and looked at the deputy chief, saying, "Chief, that's impossible; there was nothing left to do up there."

The chief was in a bind. He had to respect this transmission, so he ordered another engine to stretch a hoseline. Danny went back upstairs to demand an explanation. He found the ladder officer and asked him what was happening. "Look here, you see this dresser," the officer replied. "It's still smoldering."

The ladder company lieutenant insisted there was a problem, but nothing more than embers burned on the dresser. The second line was now in place on the top floor. Danny angrily stormed out and returned to the command post, where he later witnessed massive amounts of water coming out the windows. The staircase looked like Niagara Falls. When the officer from the second engine came back down to the street, he asked him what that was all about. The man replied that he "had a new guy that I wanted to do a little drill with."

Danny couldn't believe it. He had never seen anything like this after working in busy areas with the best in the business. It was going to be a big adjustment. What should have been an incipient-stage fire that could have been controlled by a portable fire extinguisher (a can job) turned into a major cluster. Instead of just a mattress needing to be replaced with a little cleanup, this was now a major overhaul. Water damage alone was more than what the fire had destroyed.

11

Katrina

When Danny was a lieutenant, his company received a visit from the second-in-command chief of a large urban area in Latin America. After spending a week in Danny's firehouse, the chief had asked Danny if he would like to visit his city, Guayaquil. Even though they could only communicate through translators, a strong friendship evolved. He welcomed the opportunity; it would mean a break from the pile after 9/11, funerals, and the firehouse. He told the chief he would go on one condition—if he could do some training.

When he arrived at Guayaquil, he couldn't believe how modern it was. It looked like a smaller version of his own city, but the fire department wasn't. They didn't have any of the modern equipment that firefighters in his department took for granted. Basic PPE like SCBA, bunker gear, tools, or anything else firefighters used on the fireground in New York was nonexistent.

After staying a week, he vowed to help, and thus the relationship was formed. There were many training trips, and fire trucks of all kind and PPE were sent down over the course of about 10 years.

Danny started a program where he would bring up a small group of firefighters from that city to do an advanced train-the-trainer program in a modern fire academy once a year. One afternoon around Labor Day, they were eating lunch in the cafeteria. The news was on the TV in the background and they were talking about this category five hurricane that was set to strike New Orleans. Danny paid no attention to the storm; it was so far away that it would have nothing to do with him. They wrapped up the training that day and he headed back to his city, where he was scheduled to work the next night tour. The hurricane slammed into New Orleans with a vengeance, wreaking total devastation.

When he arrived for the night tour, he headed to the hub of every firehouse, the kitchen, for a cup of coffee. This was his ritual for the past 20 years: he liked to get in at least an hour before the tour started. He noticed the news was running continuous coverage of the hurricane. It was the first time he had seen any news from the area. He couldn't believe the destruction.

He got settled in the office and noticed a fax from the division headquarters sitting in the hopper. They were looking for 300 volunteers to go to New Orleans to assist the fire department. He canvassed the guys in the kitchen, and within minutes the spots were filled. Danny sent a fax back to the division with 10 names on it. They didn't think twice about giving up their last weekend of the summer to help out. So many firefighters volunteered that they had to turn guys away. It was a chance to give back for all the help they received after 9/11.

The next morning was quite a sight, with 300 firefighters boarding a chartered plane, carrying their bunker gear like they were going on a detail to another firehouse. It was comical: guys had brought their bunker gear and helmets with them onto the plane like carry-on luggage.

When they opened the doors to the plane at the New Orleans airport, it felt like an oven. The heat and humidity was oppressive. The city was in shambles; they saw planes flipped, boats in the street, and roofs of buildings torn off. There was no power, and it seemed like Armageddon. They boarded a few buses and headed to a Catholic nursing home that was going to become base camp for the next 2 weeks.

After 9/11, the department set up an incident management team (IMT), composed of five sections: finance, logistics, operations, planning, and command. A team of roughly 40 handpicked chiefs established the first ever IMT. They realized that they needed more than their own SOPs to handle large-scale incidents. This introduced some novel ideas that revolved around the incident command system (ICS), such as establishing branches, divisions, and groups, among many other ideas. If the city were ever to experience a major catastrophe, they would have their own team available to run the operation. New Orleans was the first major deployment for the new IMT. They were going to manage the 300 firefighters assigned to the disaster.

It was a sweltering first night at the base camp, and there was nothing; no power, food, or adequate bed space. Guys and gals found any available floor space to set up camp. Danny found a spot in the sacristy of the chapel to lay his head. One of the guys raided the kitchen and put together a meal out of canned food until Logistics could set up a functioning kitchen.

The New Orleans firefighters were fighting on two fronts:

- The insane amount of fires in the city that were burning out of control: the hydrant system in the city was out of service because of the storm.
- Their own personal disasters: the firefighters all lived in the area and were trying to get their own houses back in order.

They had been on duty since the storm hit, and many were up for 48 hours without sleep. They were getting clobbered.

The next morning, the quiet was disrupted at daybreak around 0600 hours. There were still a few fires that had burned through the night that the firefighters couldn't stop because of the lack of water. The fires burned until they had nothing left to burn. A chief from the New Orleans Fire Department (NOFD) burst into the command tent and asked the incident commander for help: "The city is burning down, we need help, our guys are shot. Can you give us 75 firefighters to man the rigs?"

It was like sailors on a ship when they call "all hands on deck." The chiefs came into the bunks and said that they needed firefighters to man the apparatus. You should have seen how the guys scrambled. The New Orleans fire engines were lined up in the base camp and firefighters were jumping on them. They looked like jet planes lined up on an aircraft carrier. When an engine had five firefighters and an officer it would take off for any of the multiple fires that were burning.

Danny got on one of the first trucks out that day and was assigned to a group of houses that were burning down in the center of the city. When they arrived on scene, they felt helpless. Without a positive water source, they could do nothing. They went into an exposure on the D side with a charged hoseline. They had 1,000 gal of water to try to stave off the impending conflagration. The chauffeur/apparatus operator was desperately trying to draft from the standing flood water but wasn't having much success.

The local cell phone company had hired some milk trucks filled with water to supply their building downtown to keep their air conditioning running. At night the trucks were staged on the high ground on the edge of the city. Danny had heard about this and had an idea. He thought that if they could assign one or two of the trucks to a working fire, they could have a positive water supply. The only issue would be getting the water from the milk truck to the fire engine.

When Danny approached one of the chiefs on the IMT and presented his idea, the chief replied, "Sure, knock yourself out, give it a try."

Danny knew that there were a few guys on the team who had experience in volunteer fire departments. One lieutenant had been on one of the five department satellite engine units, which are capable of supplying large amounts of water to multiple alarms. They grabbed one of the New Orleans fire engines and set out to see how they were going to get water out of a milk truck. When they got to the milk truck, Danny sized up the situation. There were no outlets on the truck that were compatible with fire hose, so they needed a plan B.

Plan B was to attempt to draft from the top hatch of the milk truck. The chauffeur set up the engine for drafting. The rest of the guys stretched a supply line and dropped it down the hatch. The chauffeur took the steps necessary, and before you knew it water was flowing. They had successfully figured out a way to draft water from this 10,000 gal milk truck. Danny was ecstatic; he couldn't wait to get back to base camp to share his findings with the chief.

He made his way to base camp and came across his chief's boss. "Hey chief, we figured it out. We can draft from the milk trucks."

The chief smiled. "Great news, Cap, but forget about the milk trucks. I'm giving you 80 wildland tankers and you're going to be working with another supervisor from the Wildland Fire. I need to know ASAP how many you can use; make it work." Wildland tankers are synonymous with wildland tenders, and supervisors are known as division supervisors (DIV SUP).

Danny was stunned and also a bit offended. He thought to himself that maybe the chief didn't think he could handle it. Nevertheless, he now had some figuring to do. He had to do the math to determine how much water was needed on scene at any given time to have an adequate water supply. He had to consider the following:

- How much each hoseline would require, considering
    - 1¾ in. hoseline = 180 gpm
    - 2½ in. hoseline = 250 gpm
- How much water each wildland tender carries
    - Type 2 carry 2,500 gal of water
- Portable tanks hold about 2,500 gal of water
- How long it would take for each tender to dump and return to water supply (a NOFD fireboat)

His goal was to have 5,000 gal on scene at any given time. It was decided to have three task forces, each made up of the following:

- One Type 1 engine capable of pumping 1,000 gpm
- One Type 1 engine with a front suction
- Five Type 2 water tenders at 2,500 gal
- Two portable water ponds

The three task forces (designated Water Tender Task Force 1, 2, and 3) were spread out to cover the whole city. The system worked, and a task force would roll on every working fire. ("Roll" is a New Orleans term for a call to a working fire. FDNY uses "run.") The water problem was solved. Hoselines were now being charged and operating without fear of running out of water. The three task forces came together as a team very quickly. Their collective experience in the wildland was a huge asset to the operation.

The teams were in a groove, as if they had been doing this together for years. One day when Danny was out checking on the three teams, he was talking to the team leader of Task Force 1 and noticed that the man was wearing a shirt that said "Raindancers."

He asked him, "Who are the Raindancers?"

"That's the name of my company; some of the trucks here are on my team," replied the leader.

Danny said half-jokingly, "From now on you are no longer Water Tender Task Force 1, but the Raindancers."

He left the Raindancers and headed over to the next task force down on Decatur Street. He did his usual checking in. The guys were tired from the night before. He caught up with the team leader: "I see your team doesn't have a name; why's that? You know Task Force 1 calls themselves the Raindancers."

The team leader was a bit miffed. "Give me a minute—you know what, we do have a name. We're the *Water Wizards*" (fig. 11–1).

Danny laughed, "10-4, from here on out you are no longer Task Force 2, but the Water Wizards."

He had started something here and now had to see it through, so as you can guess, when he got to the third task force, he didn't even wait.

Figure 11–1. Water Wizards

"You know, I just left the other two task forces and they both have named their teams: Task Force 2 is the Water Wizards and Task Force 1 is the Raindancers. What about you guys?

The leader didn't even balk. He replied, "We're the *Water Dogs*."

What had started out as a joke actually took hold. The command staff and the dispatchers were having a blast. Instead of calling out, "Dispatch to Water Tender Task Force 1, you gotta roll," it was now "Dispatch to the Raindancers, you gotta roll."

It actually made the teams tighter because now they had an identity. Most firehouses in the country have names they go by. In FDNY, Engine 60, Ladder 17, and the 14th Battalion are known as the Green Berets. It becomes a source of pride. The Water Tender Task Forces had a generic designation, but just by the simple act of giving them a team name, they became a unit. The task forces remained in place for 2 years after the storm until the hydrant system was fully functional

Rural Water Supply Operations in the Big Easy

After two days of extremely grueling firefighting operations and the loss of numerous buildings due to lack of water, FDNY/NOFD members came up with a plan to allow us to assist in saving what was left of the city of New Orleans.

Before FDNY arrival, NOFD members were able to draft flood waters to supply hoselines. However, as the water levels dropped and numerous neighborhoods began to dry out, water supply became a real problem. The pumping station that supplies the hydrant system was flooded.

Over-the-road tank trucks were brought in to supply water. Hard sleeves were dropped through the top hatch of the tank truck and then connected to the steamer fitting on the NOFD engines. Several problems arose from this operation:

- First, the tankers were very long and had trouble maneuvering through the tight city streets.
- Second, once they were in place, you couldn't get out to refill.
- Third, many times the engines couldn't attain draft due to the length and bend of the hard sleeve.
- Fourth, the hard sleeve with a barrel strainer was only capable of drafting to within approximately 18 to 24 in. of the tank bottom and thus unable to offload approximately 2,000 of the 8,000 gal on the truck.

With multiple fires on the same or adjoining blocks, 6,000 gal didn't last very long. There needed to be a plan to use more mobile firefighting tankers to supply the needed water. However, some of the arriving tankers' hose fittings did not conform with NOFD thread. It was decided NOFD Engine 8 (The Buffalo Soldiers) would be converted into the "water supply engine." Engine 8 would run every reported fire citywide.

NOFD and FDNY members under the direction of an FDNY officer went to the NOFD shops and equipped Engine 8 with numerous combinations of fittings that allowed Engine 8 to supply or take water from any rig, including those coming from other states. Next, the incoming tankers were broken up into groups of at least six tankers. Three groups of tankers were deployed in the city along with a supply engine with a preconnected hard sleeve mounted on the front bumper. Engine 8 would run with all three groups.

The Plan

When units received a report of fire, Engine 8 along with the first two tankers would respond directly into the scene, staging just beyond the nearest intersection. The remaining tankers and the supply engine would stage one block beyond Engine 8. Upon the report of a working fire by the first-due companies, Engine 8 members would pick up the first-due engine's supply line and extend it to a 5 in. × 5 in. × (4) 2½ in. gated wye/Siamese.

The first-arriving tankers (Tanker 1 and Tanker 2) immediately supply the gate valve. Engine 8 and Engine 35 members stretch a 5 in. supply line from Engine 35 to Engine 8. Engine 8 also connects to the gate valve/Siamese. Tanker 3, 4, 5, and 6 members drop their folding portable tanks and set up the tanks in front of Engine 35. Tankers 3, 4, 5, and 6 dump their water in the tanks. Engine 35 drops the hard sleeve into the tank, attains prime, and supplies Engine 8. Once Tankers 1 and 2 are empty, they pull past Engine 8 and go to the refill site (a fireboat) to become part of the water shuttle operation.

Apparatus Positioning

Engine 8 had to leave the intersection partially open for later-arriving apparatus and at the same time leave enough room for Tankers 1 and 2 to get past when they went to refill. Parked cars became a problem several times, but the apparatus just went up on the sidewalk to make room.

The water shuttle operation was kept two blocks away to allow free movement of both later-arriving apparatus and the tankers. Engine 35 backed into the block to allow room for the portable tanks and multiple tankers dumping at the same time. Even after the pump house was up and running, water pressure was very poor. The tanker operations continued to supplement the hydrant system.

Epilogue: 9/11

Danny was annoyed with himself that he had to study once again for the upcoming captain's exam in October 2001. He really should have passed the first one years earlier. But then, he felt like he was too young to be a captain. He recalled studying half-assed, hoping to barely pass and come out at the end of the list, which would mean another 4-year wait for his turn. Lesson learned: never do anything halfway. If you're going to do something, do it 100% or don't do it at all.

He found his groove and was crushing the practice exams. With the test a month away, it was time to start his final review. His two older girls were on the school bus, while he was ready to take the youngest to preschool. He loaded up her little backpack with a lunch, juice box, and a snack. His books on the dining room table would be waiting for him when he got back. He had the entire morning to himself.

The school was about 2 miles away, and Danny liked to listen to Howard Stern on the radio in the car. He was insane but very entertaining. As they got about halfway to the school near the lake, the cohost Robin Quivers burst in on the radio: "We are getting a report that a plane has just hit the World Trade Center."

Danny thought to himself that it may be some sort of a joke. He said to his little daughter, "Not funny, Howard."

It was a perfect September morning, not a cloud in the sky. He thought it was probably a small two-seater, perhaps a pilot that had a medical issue and veered off course.

Robin came on again: "They are reporting now that a commercial jet has hit the North Tower."

He turned on the local news channel to see what the heck was happening. It was true. A jet plane had slammed into the World Trade Center. He dropped his daughter off and explained to the school staff that he was a firefighter and needed to get to the firehouse ASAP. They took her in and told him to go with

God's speed. He raced home and couldn't believe what he was seeing on TV. There was a massive hole on the north side of the North Tower.

He was scheduled to work that night, so he prepared to leave as soon as he could get his stuff together. He called his wife and told her that the kids were in school and that he was heading in early. She told him, "I heard that plane go overhead."

He got ready in about 10 minutes with the TV showing what appeared to be another plane heading for the South Tower. His mind did not register what was happening: "Is it possible that another plane was hitting the *other* tower?"

He jumped in the car and sped toward the city 50 miles away. His roller blades were in the trunk. He was getting into the city one way or another. The parkway was empty; the only people on the road were a few first responders. His old five-speed Honda Civic was pushed to the limit, with the tachometer cranking at 4,000 RPM the whole way. Listening to the news radio, his thoughts were that at most maybe there would be a localized collapse of a few floors, not a catastrophic collapse.

It's a massive event when a building collapses. He'd seen a few building collapses when he was a firefighter, with thousands of tons of brick crashing down, but never in a million years would he have expected what he just heard on the radio: "The South Tower of the World Trade Center has just collapsed. God help us."

Stunned, he could not get his head around what was happening. He met up with one of the "squads" by Yankee Stadium with an officer and a friend from days of working in Special Operations Command (SOC). He used the rig to block traffic and was able to make his way over the bridge. The city was a ghost town.

He was the first one in; his engine company had been assigned on the fifth alarm but the ladder company was still in quarters. The rest of the firefighters started arriving shortly after. When they had enough firefighters, they went out to the avenue and flagged down a city bus.

"We need to get to the WTC site." The bus driver replied, "No problem."

They gathered every bit of equipment they had and boarded the bus. They made three stops at some other firehouses before making their way down the east side of the city. It was like a scene out of a horror movie, people wandering the streets covered in dust, aimlessly trying to escape the disaster. The bus took them about six blocks south of the WTC, as close as they could get.

Danny ordered the firefighters to take all the heavy rescue tools (Hurst tool, air bags, power saws, and assorted hand tools) off the ladder truck they encountered on Rector Street. It was an eerie feeling to see the firefighters' shoes strewn about the fire truck. Danny knew that they would be encountering massive amounts of debris that would have to be moved to get to trapped civilians. He had no idea what was ahead at the site but thought that at least they should be equipped. He would be remiss if he didn't have a Handie-Talkie, but there were just none available. After they gathered up the equipment, they headed to the site a few blocks away.

As they got about a block away, they came across 90 West Street, an older office building where the basement and a few of the floors were fully involved with fire. A chief ordered him to get a hoseline in there and put out the fire. He was taken off guard. How was he supposed to put out five floors of fire with no pumper and no water? The team had no SCBA, radios, pumper, or any working hydrants. It was chaotic with fires burning everywhere. He thought about the scene as if a nuclear bomb had been dropped on the WTC and can recall thinking to himself this was like Ground Zero.

He had just been here only a week prior and remembers how great this place had looked and how they had done so much to improve the area. Now, it looked like Armageddon. When they crossed under the overpass on Liberty Street and saw the scale of the destruction, he turned to the guys and said, "Put the tools down."

He bumped into a good friend of his from the squad and asked if he was working.

The friend looked at him with a vacant stare, responding, "If I was working, I'd be dead."

The first order of business was to find out about his company. He searched the pile, and after about a half hour he found one of his firefighters. He was badly injured but was ambulatory. Danny went with his firefighter to the spot and asked him about the status of the guys.

"Joe's gone. We were talking to him one second, the building collapsed, and he disappeared."

He proceeded to tell him the story of what happened. Joe was a hardcore, tough-as-nails Vietnam vet. He told them on the way down that they were going to go slow and stick together. They all grabbed an extra air bottle and headed to the command post. The chief at the incident command post ordered them into the lobby of the Marriot Hotel.

While they were in the lobby with a few other companies and a few chiefs, the South Tower collapsed. Joe disappeared. Even though he was severely injured, his only concern was the guys, especially the probie. They told him they were banged up but okay. They threw a yellow rope to him and started digging.

Even though he was buried under tons of concrete and debris, the guys didn't give up. They continued digging by hand to get to him. They could hear him, knowing he was buried deep. The second collapse occurred, and then he was gone. The guys hung in there the whole time, surviving both collapses. The only thing that remained was the severed rope.

There was a commotion over by the North Tower. Since Danny didn't have a Handie-Talkie, he wasn't exactly sure what was happening. It turned out that there was a group of firefighters trapped in Stairway B of the north tower. Danny and his crew climbed over a mountain of debris to get to the spot where the firefighters were buried (fig. 12–1). One of the firefighters who was trapped was a friend of his who worked with him in a South Bronx ladder company.

Back when Danny was a firefighter in the squad, his lieutenant brought in a load of 5 gal spackle buckets. One night, instead of a drill, they spent a few hours taking them apart. The guys thought he was nuts, but since he was the boss,

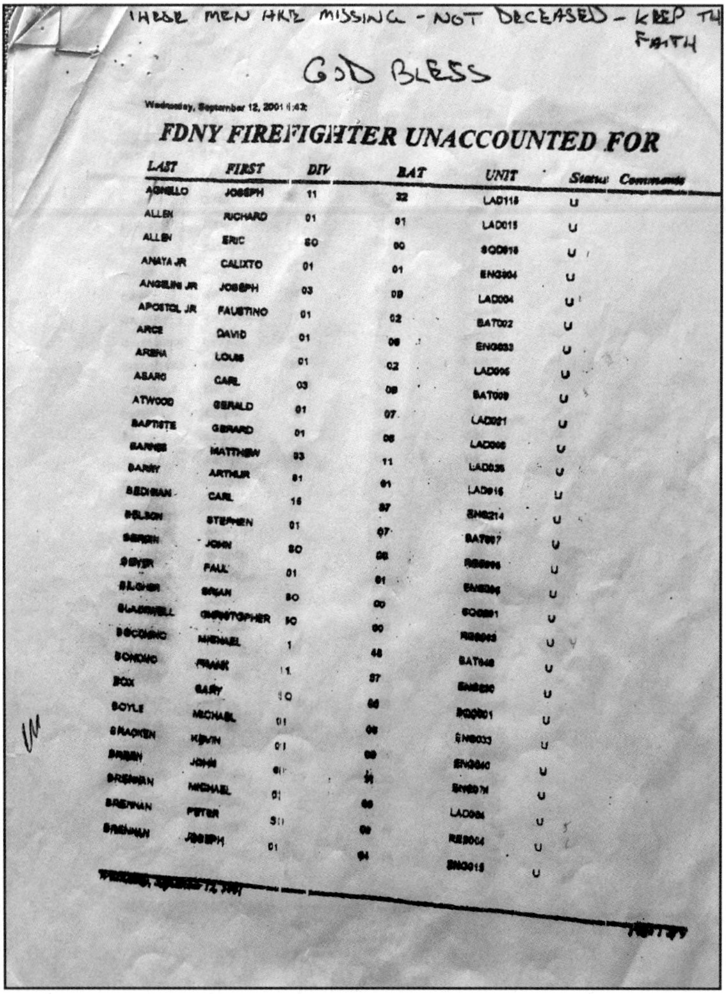

Figure 12–1. List of unaccounted for FDNY firefighters

they followed orders. Since they were going to so many building collapses, they were going to carry them on the apparatus to help remove debris.

It turned out that his lieutenant was right: they were invaluable when it came to removing debris. There was a huge line of firefighters on the pile with 5 gal spackle buckets moving tons of debris by hand. It took a while, but eventually all the firefighters who were trapped were removed, along with a few civilians and police.

So much happened that day with 110-story towers collapsing, but don't forget that 7 WTC, a 48-story office tower, collapsed as well. The whole time they were on the pile moving buckets of debris, 7 WTC was burning out of control. There was a massive fire at 5 WTC as well. It was apparent that 7 WTC was going to collapse, and it later did that afternoon. By the grace of God no one was hurt when the building collapsed.

The department had suffered a tragedy in June of that year. Three firefighters were killed when a hardware store fire turned into an explosion and collapse. It was a major event. To lose even one firefighter sets the department back on its heels, but to lose three was unthinkable. But this day, rumors began circulating at the pile that a lot of firefighters were missing. We first heard a rumor that something like 25 had perished, which was unfathomable. That number climbed throughout the day to the point where they put out a list. There were almost 400 names on the list.

How do you get your head around that number? The list looked like a captain's promotional list. They all sat down and looked at the names. Danny had worked in SOC when he was a firefighter and lieutenant and knew at least a third of the names. He went numb and was overwhelmed, not digesting what was happening. He had worked at this location when he was in college and remembered it as a great, vibrant place. Now it looked like a wasteland.

It was late; he had been there for about 12 hours and forgot to call his wife. He needed to let her know that he was all right and also to check in to see if she and the kids were okay. He needed to find a pay phone. He didn't own a cellphone, but it didn't matter because they weren't working anyway. He heard that there was a bank of pay phones in the lobby of the financial center. It would be tough to get to through all the debris, but he found a shortcut.

On West Street there was a nondescript door with no markings (fig. 12–2). He knew that if he went through that door he wouldn't have to climb over the mountain of debris. He opened the door, entering a dark hallway. It was eerily quiet, but there was a pungent odor in the hall. It was the same unique, acrid pungent smell he remembered at the pile, something he would never forget. When he walked deeper inside, the odor got stronger until it became overwhelming, smelling like death.

Figure 12–2. A door that ended up leading down a dark hallway

He had smelled it numerous times in his career, but this was more intense. He continued down a pitch-black corridor and could see a room to the right. The floor was beginning to get very slippery. There was liquid all over the floor. He got to the door, and that's when he realized that he had found the morgue. Apparently, firefighters killed in the collapse were being kept in this room, spread out on metal tables in the vast room. He scrambled to get out of there, slipping on the blood all over the floor. Shaken, he eventually found the phone bank and was able to get through to his wife.

He stayed down there until 0400 hours. They hitched a ride back to the firehouse in the back of a pickup truck. The following 6 months were a blur, a cycle of funerals, working the pile, and back to the firehouse. They recovered the last few bodies they could by March 2002.

They never found the body of his beloved comrade, who died a most noble death. He had put himself in harm's way, his only concern being for his fellow firefighters.

> *Greater love hath no man than this, that a man lay down his life for his friends.*
>
> —John 15:13

Index

About the Author

Danny Sheridan is an FDNY battalion chief with a history that's taken him all over the world in firefighter training, safety, and cause and origin as part of a nonprofit he developed and continues to lead. With his nearly 37 years with the FDNY, he served as an operations section chief, FDNY Incident Management Team member, and was a branch director during Hurricane Sandy. Danny has worked in Harlem and the Bronx for most of his career, and previously he instructed at the Rockland County (NY) Fire Academy. He is a frequent contributor to *Fire Engineering*, has a monthly column on FireEngineering.com, and hosts a podcast called *First-Due Battalion Chief*. He authored chapter 12 (Forcible Entry) for *Fire Engineering's Handbook for Firefighter I and II*.

In 2005, Danny founded Mutual Aid Americas Inc., which was in operation from 2005 until 2011, with the mission of sharing the valuable knowledge and expertise gained during his decades of experience with fire departments in Latin America. Under the Mutual Aid Americas banner, he led, conducted, and took part in numerous in-depth firefighting trainings and seminars for major firefighting corps throughout South America including Cuenca, Guayaquil, Biblián, and Monteverde, Ecuador; Guadalajara, Mexico; Valdivia and Santiago, Chile; and Lima, Peru. Mutual Aid Training Group LLC was founded in October 2011 and follows in the footsteps of Mutual Aid Americas, and Danny continues his work of giving back—and passing it on.